Low-Noise Low-Power Design for Phase-Locked Loops

Feng Zhao • Fa Foster Dai

Low-Noise Low-Power Design for Phase-Locked Loops

Multi-Phase High-Performance Oscillators

 Springer

Feng Zhao
Santa Clara
California
USA

Fa Foster Dai
Department of Electrical and Computer En
Auburn University
Auburn
Alabama
USA

ISBN 978-3-319-34370-9   ISBN 978-3-319-12200-7 (eBook)
DOI 10.1007/978-3-319-12200-7

Springer Cham Heidelberg New York Dordrecht London
© Springer International Publishing Switzerland 2015
Softcover reprint of the hardcover 1st edition 2015

Printed on acid-free paper

Springer is part of Springer Science+Business Media (www.springer.com)

Preface

This book covers unique design techniques for phase-locked-loop (PLL) frequency synthesizers, including quantization noise reduction, low power design methodology for dividers, and high-performance multi-phase clock generation techniques.

PLL is a critical building block for frequency synthesis in wired and wireless communication systems. This book first introduces fundamentals about fractional-N PLL design. It analyzes the quantization noise of $\Sigma\Delta$ modulator based fractional-N PLL. $\Sigma\Delta$ modulation is a commonly used technique for fractional spur reduction. However, quantization noise caused by $\Sigma\Delta$ modulator degrades the phase noise performance at the PLL output. The noise degradation becomes worse when the loop is nonlinear. In this book, noise reduction and cancellation techniques are presented to solve the problems associated with noise degradation due to $\Sigma\Delta$ modulation. Also analyzed is the impact of charge pump nonlinearity on phase noise performance. System-level behavioral model is developed for the analysis of charge pump nonlinearity.

For low power PLL design, a wideband integer-N PLL with power optimized dividers is presented. An intuitive, yet efficient power optimization methodology is introduced for optimizing divider power consumptions using both bipolar and CMOS technologies.

Quadrature signals are needed in image-rejection transceivers. This book presents the analytical results and design details of two quadrature voltage-controlled oscillators (QVCO) with capacitive-coupling technique. Capacitors are used to couple the two voltage-controlled oscillator (VCO) cores. The proposed capacitive-coupling QVCO (CC-QVCO) architecture has the advantage of low phase noise performance and it is free from the bi-modal oscillation, which causes the quadrature phase ambiguity. Unlike conventional quadrature-coupling scheme using active devices, CC-QVCO utilizes noiseless capacitors to form QVCO, allowing phase-shifted coupling signals that lead to reduced phase noise.

A differential Colpitts CC-QVCO with enhanced swing is proposed to offer excellent phase noise performance under low power supply. It achieves lower phase noise than its single-phase counterpart. Optimized capacitive coupling combined with source inductive enhance-swing technique achieves low power and low phase noise simultaneously. Due to the inherent phase shift in the proposed quadrature-coupling path, the problem associated with $\pm 90°$ phase ambiguity between the quadrature outputs is eliminated.

Capacitive-coupling technique can also be applied to multi-phase clock generation for improved phase noise performance. A QVCO prototype using classic NMOS cross-coupled VCO with current tail is implemented to demonstrate the advantages of capacitive-coupling comparing to other coupling techniques. The problem of phase ambiguity for this QVCO is also avoided by the intrinsic leading phase shifting through the capacitive coupling paths. Circuit implementations and measurement results of this CC-QVCO and another class-C mode top-series QVCO (TS-QVCO) used as a comparison are presented with details. This book is organized as follows:

Chapter 1 introduces several critical techniques for high performance PLL designs.

Chapter 2 explores several quantization reduction techniques for fractional-N PLL designs.

Chapter 3 describes a wideband PLL with power optimization methodology for low power divider designs.

Chapter 4 presents a capacitive-coupling technique for multi-phase clock generation using NMOS VCO with current tail. The proposed structure demonstrates excellent phase noise and accurate quadrature phases over a wide frequency range.

Chapter 5 discusses the design of a capacitive-coupling quadrature VCO with emphasis on phase noise reduction and elimination of phase ambiguity under a super low supply voltage.

Chapter 6 provides a summary for this book.

This book is mainly based on the Ph.D research work of the first author under the supervision of the second author. The authors would like to thank the team members at the RFIC Design Laboratory at Auburn University who have provided supports and helps in one way or another throughout the development of this book. In particular, we would like to express our special gratitude to Professor Bogdan M. Wilamowski, Professor Stanley J. Reeves, and Professor Guofu Niu for valuable comments on revising the manuscript. We would like to thank Professor David Irwin and Professor Richard C. Jaeger for their guidance and discussions.

All of the work was made possible by the love and encouragement of our families. Feng Zhao would like to thank his parents and family members for their supports. He would like to express his cordial appreciation to his wife and his son for their continuous love and support.

Contents

List of Figures

List of Tables

Chapter 1
Introduction

1.1 Phase-Locked Loop (PLL) Frequency Synthesizer

Phase-locked loop (PLL) circuits are widely used to generate a precise frequency signal from a very high precision reference signal. It has wide application in wired and wireless communications systems to provide accurate carrier signals that is phase aligned with the incoming high-precision reference clock signal. The voltage-controlled oscillator (VCO) signal is divided by a programmable loop division ratio N and compared with the high-precision reference by the phase frequency detector (PFD). Then, the error signal is fed into a charge pump to transform the phase error detected by the PFD into current pulses. The pulses are filtered by the loop filter and then the filtered voltage is used to control the VCO to produce a stabilized output. The negative feedback mechanism around the PLL loop results in the generation of a tunable and stable output signal at the desired frequency.

Two types of PLL structures with negative feedback loop, integer-N and fractional-N, have been adopted for frequency synthesis. Integer-N PLL, as shown in Fig. 1.1a, provides an output frequency that equals to N times of the reference frequency, where N is the divider ratio. However, this architecture limits the frequency resolution of the PLL to its PFD comparison frequency. Thus, an integer-N PLL has to use low reference frequency to achieve fine frequency tuning step, which leads to a large loop division ratio N and magnifies the in-band phase noise as it magnifies the reference frequency.

Another architecture called fractional-N PLL has become increasingly popular since its invention, because it can achieve fine frequency resolution with high reference frequency. Unlike the integer-N PLL, a fractional-N PLL allows a division ratio of fractional number by using a control module to dynamically switch the divide ratio between different integer numbers so that the time-term average of the division ratio is a fractional number. The principle of the fractional-N PLL is therefore a result of time averaging, since only integer-N division ratio can be implemented with clock edge triggered digital logic circuits. A fractional-N PLL can achieve a frequency step much smaller than its reference, while maintains high reference frequency. As a result, it can achieve low phase noise performance.

© Springer International Publishing Switzerland 2015

F. Zhao, F. F. Dai, *Low-Noise Low-Power Design for Phase-Locked Loops,*

DOI 10.1007/978-3-319-12200-7_1

Fig. 1.1 **a** Integer-N PLL
and **b** ΣΔ modulator-based
fractional-N PLL

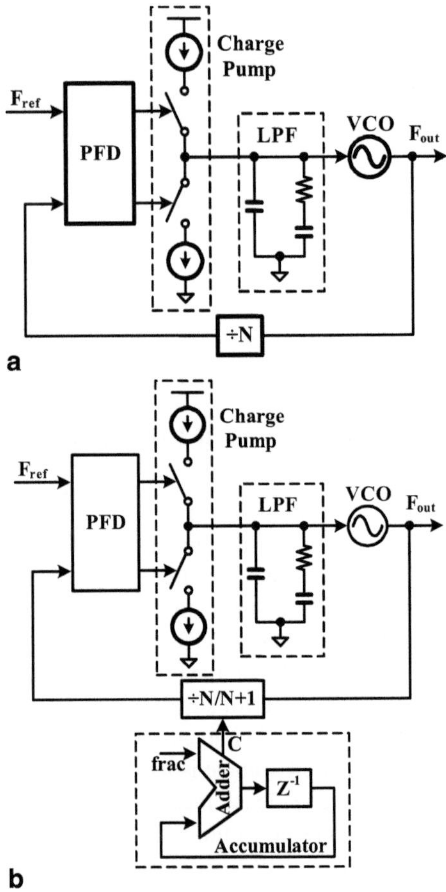

Conventional fractional-N PLL contains an accumulator or ΣΔ modulator as the fractional control module to dynamically control the divide factor [1]. Figure 1.1b shows an accumulator-based fractional-N PLL. As an example, as long as the content of the accumulator is lower than its maximum value, there will be no overflow from the adder and the loop keeps diving by N; the frequency divider is divided by $N+1$ every time the accumulator overflows. For example, with a 2-bit adder and fractional accumulator input of 1, the adder overflows every four reference clock cycles and the division ratio is thus given by

$$m = \frac{[3 \times N + (N+1)]}{4} = N + 0.25 \tag{1.1}$$

Assuming that the PLL is locked and the average voltage on the loop filter becomes zero under steady-state conditions. Figure 1.2 illustrates the resulting steady-state waveforms produced by the PLL with $N=4$ and $m=4.25$. It is evident that while the integrated current over four reference periods is zero on average, the residue value at the

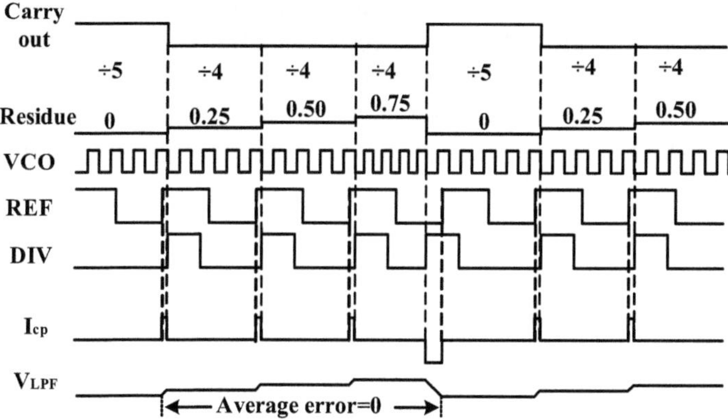

Fig. 1.2 Classic accumulator-based fractional-N PLL example waveforms for $n=4.25$

output of the accumulator instantaneously varies with time in a periodic manner. This periodicity leads to the fractional spur that plagues the classical fractional-N approach.

Then, $\Sigma\Delta$ modulator-based fractional-N PLL was proposed to address the problem associated with fractional spurs [2]. It randomizes the spur and shapes the quantization noise to high frequency band which can be attenuated by the loop. However, the instantaneous phase error still exists at the input of PFD since the $\Sigma\Delta$ fractional-N PLL still uses long-term averaging effect to achieve fractional frequency control. In order to lower the noise degradation caused by the $\Sigma\Delta$ modulator, a phase error compensation mechanism is therefore desirable to achieve the goal of wide-bandwidth modulation through fractional-N PLL.

In an attempt to reduce the noise caused by $\Sigma\Delta$ modulators, a compensating pulse-amplitude-modulated (PAM) current can be injected into the loop filter to compensate the phase error. The current is generated from a current digital-to-analog converter (DAC) with fixed pulse width. This technique can be applied to accumulator-based and $\Sigma\Delta$ modulator-based fractional-N PLL [1, 3]. The compensation technique based on PAM requires high-precision DAC current generator to completely compensate the phase error. It needs more complicated control technique such as dynamical element match (DEM) to reduce the mismatch of DAC current generator. PFD/DAC can be embedded to improve the compensation accuracy [1]; however, it can only be applied to first order $\Sigma\Delta$ modulator or modulators with a phase error less than one VCO period. Therefore, it is desirable to develop a PLL system that is able to suppress the quantization noise caused by high order $\Sigma\Delta$ modulators. A concept for high-order $\Sigma\Delta$ modulator noise cancellation is proposed in this book.

1.2   Generation of Quadrature Phase Signal

Single chip implementation of wireless transceivers gain popularity as the fabrication process is gradually migrating to smaller feature size, allowing complicated systems to be integrated with the radio frequency (RF) parts. Among many building blocks for an RF transceiver, clock signal generation is indispensable to provide clean and accurate carrier reference. Integer-N or fractional-N PLL frequency synthesizer is usually utilized to produce such high performance clock signals due to its low power consumption and accurate frequency synthesis. Inductor–Capacitor (LC) VCO plays a very vital role in providing clean clock signals because of its low power and low noise performance.

Many RF transceivers usually employ complex signal modulation and demodulation schemes because of the potential to carry more information in a limited bandwidth. Figure 1.3 shows a radar transceiver system that is able to provide gain for the signal at frequency of $f_{LO}+f_{IF}$ while reject the image signal at frequency of $f_{LO}-f_{IF}$ [4]. It requires a quadrature local oscillator (LO) signalto upconvert the baseband signal into RF. On the receiver side, the 10-MHz clock signals are also of quadrature type. Two possible phase relationships exist for quadrature output, i.e., $+90°$ and $-90°$; but it is desirable to maintain the phase relationships in one of the two possible forms since the wrong mode will amplify the image signal instead of the wanted frequency signal. It is entirely possible to embed automatic detecting circuits to find the right phase and then select the right LO signal, but it requires additional circuit. Moreover, the quadrature accuracy directly affects the signal quality in the transmitted or received signal. Therefore, a quadrature signal generation mechanism that is able to provide deterministic and accurate quadrature outputs is essential for image-rejection transceivers.

Several techniques can be employed to produce quadrature signals [5–8], i.e., (1) a voltage-controlled oscillator (VCO) with a doubled frequency followed by a divide-by-two circuit; (2) a poly-phase filter; (3) a quadrature VCO (QVCO). The first method requires a VCO operating at twice of the desired frequency which consumes more power because of the additional power consumed divide-by-two circuit,

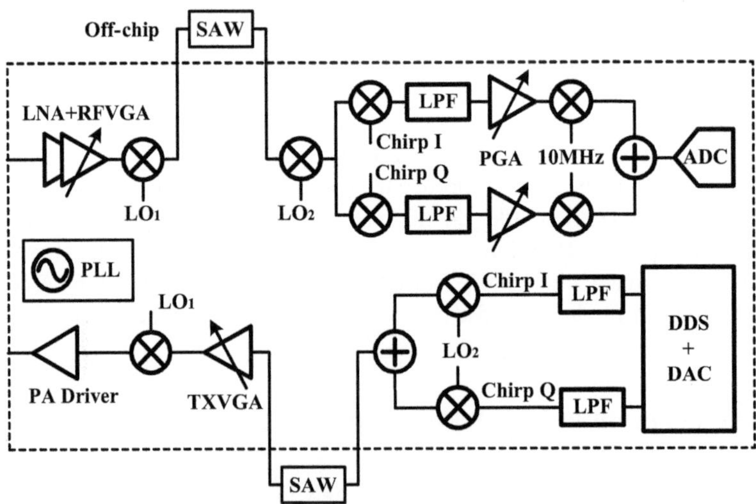

Fig. 1.3 A radar transceiver with image rejection capability

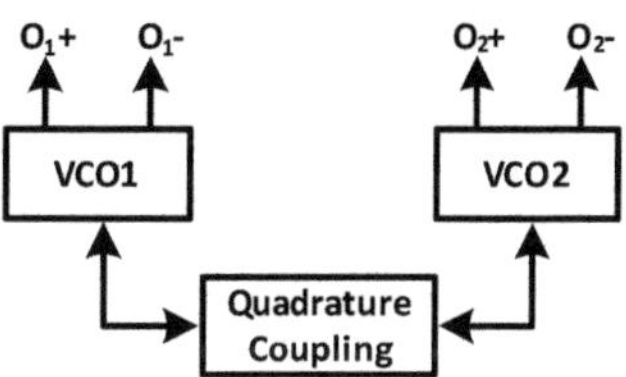

Fig. 1.4 Typical QVCO structure

and this is especially true at high frequency. The poly-phase filter is not so popular because it is a narrow-band technique with large loss. Compared with the first two techniques, QVCO comprises two VCO cores coupled with each other and can take advantage of the low power consumption. In addition, its high-voltage swing eases the design of the prescaler and the mixer. The coupling mechanism for a QVCO can be implemented using active devices or passive components like inductors, transformers, and capacitors. One popular QVCO implementation is coupled with parallel transistors due to its simplicity and low cost of area [9]. This coupling technique, however, suffers from a trade-off between phase noise and phase accuracy because the coupling needs to be strong enough to provide decent phase accuracy, which degrades the quality factor of LC tank and phase noise performance [8, 10]. Also extra power consumption is required to properly bias the coupling transistors. In order to improve the phase noise performance, transistors in series can be placed at the top or bottom of the main amplifying transistors [8, 11]; however, the parasitic capacitance introduced by the coupling transistors will reduce the frequency tuning range and the voltage headroom for the signal output is also decreased. Moreover, extra power consumption is required to maintain the signal amplitude since the coupling strength required to maintain phase accuracy lowers the signal swing. Another disadvantage of the QVCO coupling using active devices, especially with parallel transistors, is the noise degradation resulted from the current noise introduced by the coupling transistors.

A typical QVCO usually consist of two VCO cores and quadrature coupling devices, as shown Fig. 1.4. To provide the same output amplitude and oscillation frequency, the two VCO cores should have the same structure and device size. Either active device or passive components is indispensable to form the quadrature coupling between the two VCO cores such that the output can produce quadrature signals. It is obvious that the two outputs O_1 and O_2 are symmetric if the quadrature-coupling block is symmetric. Therefore, the problem of phase ambiguity, meaning that the output phase relations can be $+90°$ or $-90°$, may exist in the above QVCO structure. Fortunately, this problem can be addressed by introducing a phase delay in the quadrature-coupling path.

1.3   Prior Art of QVCO Structures

Figure 1.5 shows six types of VCO cores used in the prior QVCO topologies. Figure 1.5a shows the VCO cores used in parallel-coupling QVCO [9]. This structure is simple but suffers from the noise degradation of the active quadrature-

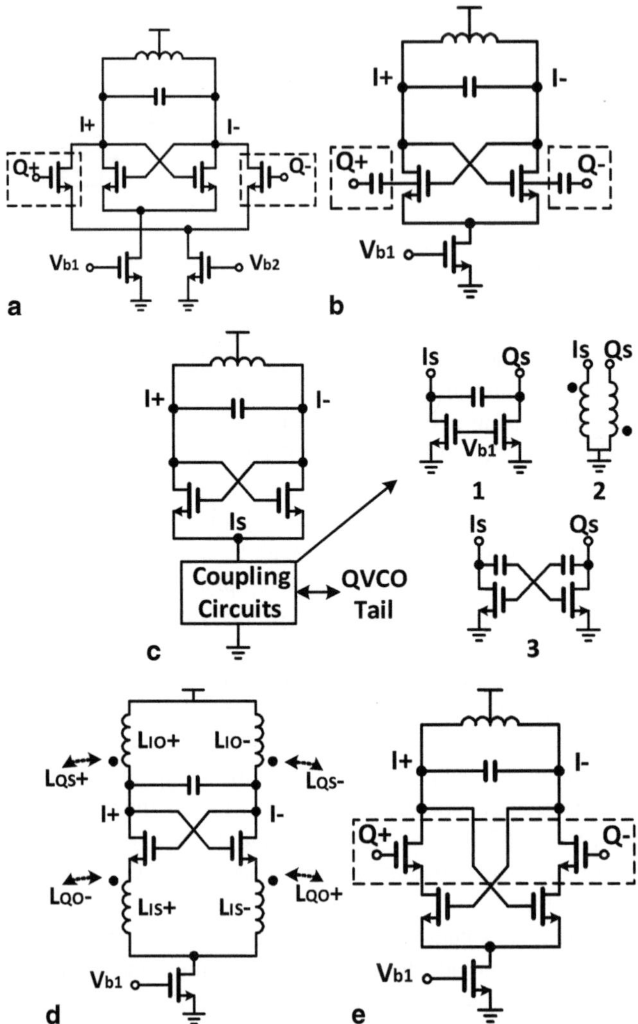

Fig. 1.5 Prototype circuits of VCO cores and coupling circuit for QVCO **a** parallel coupling [9], **b** back-gate coupling [12], **c** 2nd-harmonic coupling [5, 13, 14], **d** transformer coupling [15], and **e** top-series coupling [16]. Components in *dashed boxes* are used for quadrature coupling. The connections at Q stage are similar to I stage with I+ coupled to Q− and I− coupled to Q+. For simplicity, the bias circuitry is not shown

coupling devices. In addition, the phase delay is limited to the intrinsic delay resulted from the parasitics and thus cannot successfully avoid bi-modal oscillation.

Back-gate coupling: Fig. 1.5b shows the VCO core for QVCO with back-gate coupling [12]. This technique features the advantages of compact design and low power consumption by sharing the amplifier transistors with quadrature-coupling path. However, it requires triple-well CMOS process and is prone to the possibility of forward biasing the intrinsic bulk-substrate diode. Similar to parallel-coupling technique, it also suffers from limited phase delay in the quadrature-coupling path, which may lead to bi-modal oscillation.

Second-harmonic coupling: this technique uses the second harmonic waveform to form the coupling between the two VCO cores. Three coupling examples [5, 13, 14] with this technique are shown in Fig. 1.5c. A QVCO design using this technique consumes very small area if using capacitive coupling. However, the problem of phase ambiguity requires additional circuit to ensure correct directivity.

Transformer coupling: a VCO core used for QVCO with transformer coupling [15] is shown in Fig. 1.5d. The noise source for quadrature-coupling has been eliminated in this structure. The area cost of this QVCO does not increase much because the transformer only occupies a little extra metal area. Another advantage of this structure is that the phase delay in the quadrature-coupling path can be larger than the parallel coupled QVCO since the quadrature-coupling signal should go through a cascode transistor before reaching the LC tank. The remaining challenging for such a QVCO design is the difficulty of constructing a proper transformer model.

Top series coupling: Fig. 1.5e shows the VCO cores for QVCO with top series coupling [11, 16]. The noise contribution from the active devices in quadrature-coupling path can be reduced by utilizing top or bottom series coupling. The active coupling device is in cascode form and its noise can be degenerated by the bottom transistor. One advantage of this structure is its capability of rejecting the unwanted oscillation mode. But the voltage headroom is reduced because of the series transistors. Moreover, its phase noise performance is sensitive to bias current and temperature change.

The resonant frequency of a QVCO varies with the coupling strength and this feature can be utilized to achieve wide-band frequency tuning range [17, 18]; but the power consumption is much higher than the classic QVCO structures. As a result, this structure is not so popular for quadrature generation.

1.4   Analysis of QVCO for Deterministic Quadrature Outputs

As mentioned, the QVCO outputs can be ambiguous if not designed properly, especially under the influence of PVT variations. Directivity circuits, such as a ring of transistors [5], can be used to help produce correct output phases. Phase delay can also be introduced in the quadrature-coupling path to avoid the problem of phase ambiguity, or bi-modal oscillation [5, 19–25]. Most phase-shifting circuits achieve a phase delay much less than the optimum value of $90°$; but those phase shifters can at least provide some safe margin to avoid the problem of bi-modal oscillation.

Five types of circuit with VCO cores for QVCO are shown in Fig. 1.6a–1.6e. Cascode topology as shown in (1.6a) with a phase delay of $20°$ is proposed to move the QVCO operation away from the unstable boundary, therefore giving sufficient phase margin to avoid bi-modal oscillation [19]. The phase shift is introduced by a pole at high frequency which can be easily found from the following equivalent transconductance:

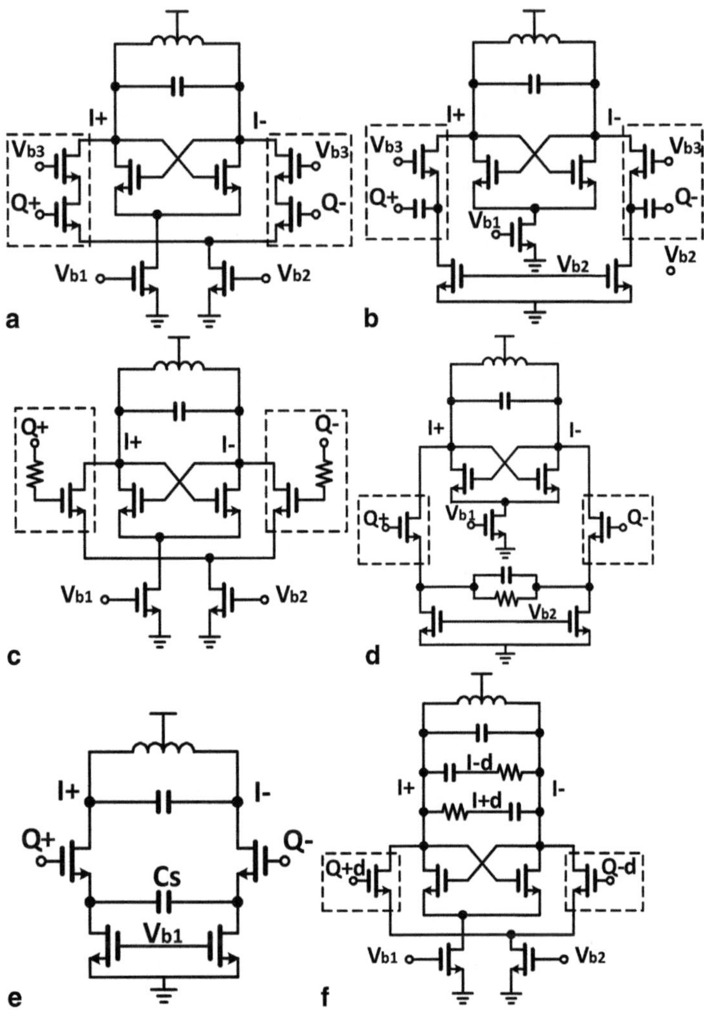

Fig. 1.6 Quadrature phase directivity circuits for QVCO **a** cascode transistor for quadrature coupling [16], **b** source coupling [19], **c** resistor-based parallel coupling [21], **d** source degenerated quadrature coupling [20, 23, 24], **e** capacitive source degeneration VCO core [17], and **f** RC polyphase filter for 90° phase shift [18]. For simplicity, the bias circuitry is not shown

$$G_{m,cascode} = \frac{g_{m1}}{1 + \dfrac{sC_P}{g_{m2}}} \tag{1.2}$$

where C_P is the parasitic capacitance or artificially introduced capacitor at the source terminal of the cascode transistor, and $g_{m}2$ is the transconductance of the cascode transistor.

The phase shifter shown in Fig. 1.6b and c can provide phase delay for stable operation of QVCO without bi-modal oscillation but both of them suffer from the noise degradation. The quality factor of the LC tank (Fig. 1.6b) is decreased because the source input impedance of $1/g_m$ will load the resonant tank; while the series resistors in the quadrature-coupling path add to the output noise.

Another type of phase shifter uses RC source degeneration network to provide phase delay [23, 26, 27], as shown in Fig. 1.6d. This structure is advantageous over cascode phase shifter because it does not suffer from the problem of degraded voltage headroom and can be embedded into the main VCO cores as shown in Fig. 1.6e [20]. The effective transconductance of the quadrature-coupling circuits for Fig. 1.6d and e are

$$G_{m,RC\ deg} = \frac{g_{m1}(1+sR_SC_S)}{1+g_{m1}R_S/2+sR_SC_S} \tag{1.3}$$

$$G_{m,C\ deg} = \frac{g_{m1}sC_S}{g_{m1}/2+sC_S} \tag{1.4}$$

Usually, the phase shifters mentioned above only achieve a phase shift around $45°$ in practical QVCO design and it is still far away from the optimum condition of $90°$. Mirzaei [21] suggested RC poly phase shifter to produce the $90°$ phase shift for optimum operation of QVCO as shown in Fig. 1.6f. A phase shift of $72°$ has been achieved due to the load effect in the real implementation according to the publication. However, the quality factor of the LC tank can be easily degraded by the RC poly-phase filter, especially when a high-quality tank is required for low noise VCO design.

1.5 Outline of the Book

This book introduces several critical techniques for high-performance PLL design. First, Chap. 2 explores several quantization reduction techniques for fractional-N PLL. Nonlinearity analyses of four types of popular $\Sigma\Delta$ modulator structures and simple noise reduction technique are discussed. An innovative technique for high-order $\Sigma\Delta$ modulator noise cancellation is proposed for fractional-N PLL.

Chapter 3 describes a wideband integer-N PLL with 4.8–6.8 GHz output frequency range. The details about the VCO, multi-modulus divider, and bandgap reference designs are presented. A power optimization methodology is also developed for low-power divider design.

In Chap. 4, capacitive-coupling technique, combined with intrinsic phase shifter, is applied to classic NMOS VCO with current tail for multiphase clock generation. The proposed structure demonstrates excellent phase noise and phase error performance over a wide frequency range. Implementation and measurement results are given to show the robustness of the proposed prototype.

Chapter 5 focuses on the topic of capacitive-coupling quadrature VCO, particularly on phase noise reduction and elimination of phase ambiguity under super low supply voltage. This chapter aims to develop a differential Colpitts QVCO with enhanced swing technique and capacitive quadrature-coupling mechanism for low phase noise performance under 0.6-V power supply. Measured results are also given to demonstrate the proposed techniques.

Conclusions are drawn in Chap. 6 with suggestions for future research topics.

References

1. S. E. Meninger and M. H. Perrott, "A 1-MHZ Bandwidth 3.6-GHz 0.18-um CMOS Fractional-N Synthesizer Utilizing a Hybrid PFD/DAC Structure for Reduced Broadband Phase Noise," IEEE J. Solid-State Circuits, vol. 41, pp. 966–980, Apr. 2006.
2. T. A. D. Riley, M. A. Copeland, T. A. Kwasniewski, "Delta-sigma modulation in fractional-N frequency synthesis," IEEE J. Solid-State Circuits, vol. 28, no. 5, pp. 553–559, May. 1993.
3. A. Swaminathan, K. J. Wang, and I. Galton, "A wide-bandwidth 2.4 GHz ISM band fractional-N PLL with adaptive phase noise cancellation," IEEE J. Solid-State Circuits, vol. 42, no. 12, pp. 2639–2649, Dec. 2007.
4. J. Yu, F. Zhao, J. Cali, D. Ma, X. Geng, F. F. Dai, J. D. Irwin, A. Aklian, "A single chip X-band chirp radar MMIC with stretch processing," in Proc. IEEE Custom Integr. Circuits Conf., Sep. 2012, pp. 1–4.
5. S. L. J. Gierkink, S. Levantino, R. C. Frye, C. Samori, and V. Boccuzzi, "A low phase-noise 5-GHz CMOS quadrature VCO using superharmonic coupling," IEEE J. Solid-State Circuits, vol. 38, no. 7, pp. 1148–1154, Jul. 2003.
6. X. Li, S. shekhar, and D. J. Allstot, "Gm-boosted common-gate LNA and differential Colpitts VCO/QVCO in 0.18-μm CMOS," IEEE J. Solid-State Circuits, vol. 40, no. 12, pp. 2609–2619, Dec. 2005.
7. J. Crols and M. Steyaert, "A fully integrated 900 MHz CMOS double quadrature downconverter," in IEEE Int. Solid-State Circuits Conf. Dig. Tech. Papers, Feb. 1995, pp. 136–137.
8. P. Andreani, "A 2 GHz, 17 % tuning range quadrature CMOS VCO with high figure-of-merit and 0.6° phase error," in Proc. European Solid-State Circuits Conf., Aug. 2002, pp. 815–818.
9. A. Rofougaran, J. Rael, M. Rofougaran, and A. Abidi, "A 900 MHz CMOS LC-oscillator with quadrature outputs," in IEEE Int. Solid-State Circuits Conf. Dig. Tech. Papers, Feb. 1996, pp. 392–393.
10. T. Liu, "A 6.5-GHz monolithic CMOS voltage-controlled oscillator," in IEEE Int. Solid-State Circuits Conf. Dig. Tech. Papers, Feb. 1999, pp. 404–405.
11. P. Andreani and X. Wang, "On the phase-noise and phase-error performances of multiphase LC CMOS VCOs," IEEE J. Solid-State Circuits, vol. 39, no. 11, pp. 1883–1893, Nov. 2004.
12. H. R. Kim, C. Y. Cha, S. M. Oh, M. S. Yang, and S. G. Lee, "A very-low power quadrature VCO with back-gate coupling," IEEE J. Solid-State Circuits, vol. 39, no. 6, pp. 952–955, Apr. 2004.
13. B. Soltanian and P. Kinget, "A low phase noise quadrature LC VCO using capacitive common-source coupling," in Proc. European Solid-State Circuits Conf., Sep. 2006, pp. 436–439.

14. D. Guermandi, P. Tortori, E. Franchi, and A. Gnudi, "A 0.83-2.5-GHz continuously tunable quadrature VCO," IEEE J. Solid-State Circuits, vol. 40, no. 12, pp. 2620–2627, Dec. 2005.

15. A. W. L. Ng and H. C. Luong, "A 1-V 17-GHz 5-mW CMOS quadrature VCO based on transformer coupling," IEEE J. Solid-State Circuits, vol. 42, no. 9, pp. 1933–1941, Sep. 2007.

16. P. Andreani, A. Bonfanti, L. Romano, and C. Samori, "Analysis and design of a 1.8-GHz CMOS LC quadrature VCO," IEEE J. Solid-State Circuits, vol. 37, no. 12, pp. 1737–1747, Dec. 2002.

17. G. Cusmai, M. Repossi, G. Albasini, A. Mazzanti, and F. Svelto, "A magnetically tuned quadrature oscillator," IEEE J. Solid-State Circuits, vol. 42, no. 12, pp. 2870–2877, Dec. 2007.

18. X. Geng and F. F. Dai, "An 8.7–13.8 GHz transformer-coupled varactor-less quadrature current-controlled oscillator in 0.18 μm SiGe BiCMOS technology," IEEE J. Solid-State Circuits, vol. 45, no. 9, pp. 1669–1677, Sep. 2010.

19. S. Li, I. Kipnis, and M. Ismail, "A 10-GHz CMOS quadrature LC-VCO for multirate optical applications," IEEE J. Solid-State Circuits, vol. 38, no. 10, pp. 1626–1634, Oct. 2003.

20. H. Tong, S. Cheng, Y. Lo, A. I. Karsilayan, and J. Silva-Martinez, "An LC quadrature VCO using capacitive source degeneration coupling to eliminate bi-modal oscillation," IEEE Trans. Circuits Syst. I, Reg. Papers, vol. 59, no. 10, pp. 1–9, Oct. 2012.

21. A. Mirzaei, M. E. Heidari, R. Bagheri, S. Chehrazi, and A. A. Abidi, "The quadrature LC oscillator: a complete portrait based on injection locking," IEEE J. Solid-State Circuits, vol. 42, no. 9, pp. 1916–1932, Sep. 2007.

22. J. van der Tang, P. van de Ven, D. Kasperkoviz, and A. van Roermund, "Analysis and design of an optimally coupled 5-GHz quadrature LC oscillator," IEEE J. Solid-State Circuits, vol. 37, no. 5, pp. 657–661, May 2002.

23. D. Huang, W. Li, J. Zhou, N. Li, and J. Chen, "A frequency synthesizer with optimally coupled QVCO and harmonic-rejection SSB mixer for multi-standard wireless receiver," IEEE J. Solid-State Circuits, vol. 46, no. 6, pp. 1307–1320, June 2011.

24. M. Valla, G. Montagna, R. Castello, R. Tonietto, and I. Bietti, "A 72-mW CMOS 802.11 A direct conversion front-end with 3.5-dB NF and 200-kHz 1/f noise corner," IEEE J. Solid-State Circuits, vol. 40, no. 4, pp. 970–977, Apr. 2005.

25. P. van de Ven, J. van der Tang, D. Kasperkovitz, A. van Roermund, "An optimally coupled 5-GHz quadrature LC oscillator," in Symp. VLSI Circuits, 2001, pp. 115–118.

26. D. Leenaerts, C. Dijkmans, and M. Thompson, "An 0.18 μm CMOS 2.45 GHz low-power quadratureVCO with 15 % tuning range," in IEEE Radio Frequency Integrated Circuits (RFIC) Symp. Dig. Papers, 2002, pp. 67–70.

27. B. Razavi, RF Microelectronics Second Edition, New Jersey, Prentice Hall, 2011.

Chapter 2
Analysis of Quantization Noise Reduction Techniques for Fractional-N PLL

2.1 Background

High performance phase locked-loops (PLL) are widely used in wireless communication systems to provide low noise local oscillation (LO) signals for digital modulation and demodulation. It has large applications in electronic devices such as cell phones, remote control devices, laptops, and alarm systems. Usually a simple integer-N PLL consists of phase-frequency detector (PFD), charge pump, loop filter, VCO, and dividers. One drawback of integer-N PLL is that its frequency resolution is limited to the reference frequency. Thus, an integer-N PLL has to use low reference frequency to achive fine frequency tuning step, which leads to large loop division ratio and degraded phase noise. Unlike integer-N PLL, fractional-N PLL can achieve a frequency step much smaller than its reference and still maintain reasonably high reference frequency, which is the key for achieving the low phase noise performance. However, the fractional control module used in a fractional-N PLL produces quantization noise and spurs at the PLL output, which deteriorates the spectral purity of the synthesized signals.

A classic fractional-N PLL usually contains an accumulator or sigma-delta ($\Sigma\Delta$) modulator as the fractional control module to dynamically control the divider ratio. The instantaneous division ratio of the divider can only be an integer number, but its long-term average of the divide ratio is $N+\alpha$, whereas α is a fractional number. Therefore, the instantaneous phase error appearing at the input of the PFD is not always zero. This phase error modulates the tuning line of a VCO and thus creates spurious tones at the PLL output. The loop bandwidth can be reduced to filter out the quantization noise and spurs resulted from the fractional control module. However, it is highly desirable to increase the loop bandwidth of a PLL to remove the VCO noise and to speed up the lock-in time for applications that requires fast switching speed, such as Bluetooth. In this chapter, phase error compensation techniques are developed to help address these problems.

Accumulator-based fractional-N PLL structure is simple but it has very large fractional spurs at the PLL output. Therefore, $\Sigma\Delta$ modulator structures have been proposed to implement the fractional-N frequency synthesis [1, 2, 3]. Quantization

© Springer International Publishing Switzerland 2015

F. Zhao, F. F. Dai, *Low-Noise Low-Power Design for Phase-Locked Loops,*

DOI 10.1007/978-3-319-12200-7_2

noise generated at the output of a $\Sigma\Delta$ modulator will be pushed to high frequency offset which can be filtered by the low-pass loop filter of the PLL. The higher the order a $\Sigma\Delta$ modulator is, the better the noise shaping effect will be. Even though the PLL loop can provide some filtering effect to the high-pass shaped quantization noise, the filtered noise may still dominate the out-of-band noise, especially for higher order $\Sigma\Delta$ modulator. In addition, the nonlinearity of the PLL, mainly caused by the nonlinear transfer function of the charge pump, will fold the shaped noise into low frequency offset [4, 5]. In this case, the loop filter cannot filter the noise folded back into the low frequency offset. The in-band noise can become worse when a very nonlinear charge pump is used. A charge pump linearization technique is proposed to improve the phase noise performance [5].

However, the abovementioned charge pump linearization technique does not remove the quantization noise source. Thus, to reduce the fractional control module noise, several noise cancelling techniques have also been proposed for fractional-N PLL. A pulsed amplitude-modulated current can be injected into the loop filter to compensate the phase error [6, 7]. Figure 2.1 shows such a PLL system with the quantization noise cancelling technique. The current is generated from a current DAC with fixed pulse width. The quantization noise existing at the output of a $\Sigma\Delta$ modulator can be compensated with an opposite current pulse. But the mismatch between the phase error and the compensation DAC may lead to inadequate cancellation. The minimum achievable noise is usually limited by the DAC resolution and mismatch between the forward path through DAC and the feedback path through PFD.

Another noise compensation technique for an accumulator-based fractional-N PLL [8] achieves better noise cancellation results by using PFD/DAC due to its embedded charge pump in the compensation path. However, the proposed technique can only be used to for accumulator or first order $\Sigma\Delta$ modulator-based PLL since

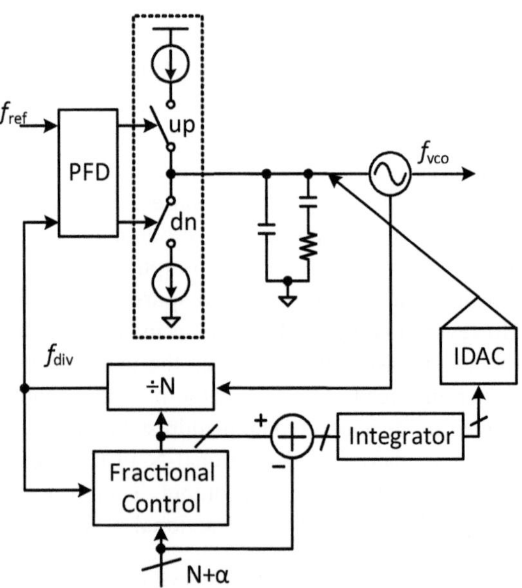

Fig. 2.1 System diagram of fractional-N PLL with quantization noise cancelling

this technique only compensates a phase error within 0–1 VCO period. However, ΣΔ modulators with orders of two or three have an accumulated phase error in the range of −2 to +2 VCO period. It is highly desirable to develop structure that is able to compensate phase error larger than one VCO period.

This chapter will analyze several different ΣΔ modulator structures and the properties of their quantization noise. Section 2.2 introduces different types of ΣΔ modulator structures and their noise-shaping effects. Noise degradation resulted from loop nonlinearity and corresponding model is discussed for ΣΔ modulator-based PLL. A noise cancelling technique for higher-order fractional-N PLL with its implementation details is described in Sect. 2.3. Finally, conclusion is given to end this chapter.

2.2 ΣΔ Modulators and Noise Folding from Nonlinearity

2.2.1 Introduction of Different ΣΔ Modulators

In order to avoid the spurious tone in the PLL phase noise spectrum, the ΣΔ modulator is usually of orders equal or higher than two. However, the modulators with orders higher than three are not so popular since their out-of-band noise may be much higher than the VCO noise and cannot be efficiently filtered by the loop. On the other hand, higher order will increase the hardware complexity of digital implementation. Therefore, the most popular ΣΔ modulators are of second or third order. MASH1-1, as shown in Fig. 2.2a, is a very classic second order ΣΔ modulator, which consists of two cascade accumulators [9]. It provides second-order noise shaping for the quantization noise with a noise transfer function of $\left(1-z^{-1}\right)^2$.

Figure 2.2b shows a third-order MASH1-1-1 ΣΔ modulator. It is simple and unconditionally stable because it does not have feedback path [10]. The overflow at the output of the accumulator is usually of one bit, i.e., either 0 or 1, so the output of MASH1-1-1 is in the range of −3 to +4 while that of MASH1-1 is in the range of −1 to +2. The MASH structure is suitable for very high clock frequencies because of its pipeline operation.

Another popular third order ΣΔ modulator is single-stage multiple feedforward (SSMF) structure as shown in Fig. 2.2c. The coefficient a, b, c in the figure can be Riley-(2, 1, 0.25) or Rhee-(2, 1.5, 0.5), but with different noise transfer function expressed as [1, 2]:

$$H_{\text{Riley}}(z) = \frac{(1-z^{-1})^3}{1-z^{-1}+0.25z^{-3}} \tag{2.1}$$

$$H_{\text{Rhee}}(z) = \frac{(1-z^{-1})^3}{1-z^{-1}+0.5z^{-2}} \tag{2.2}$$

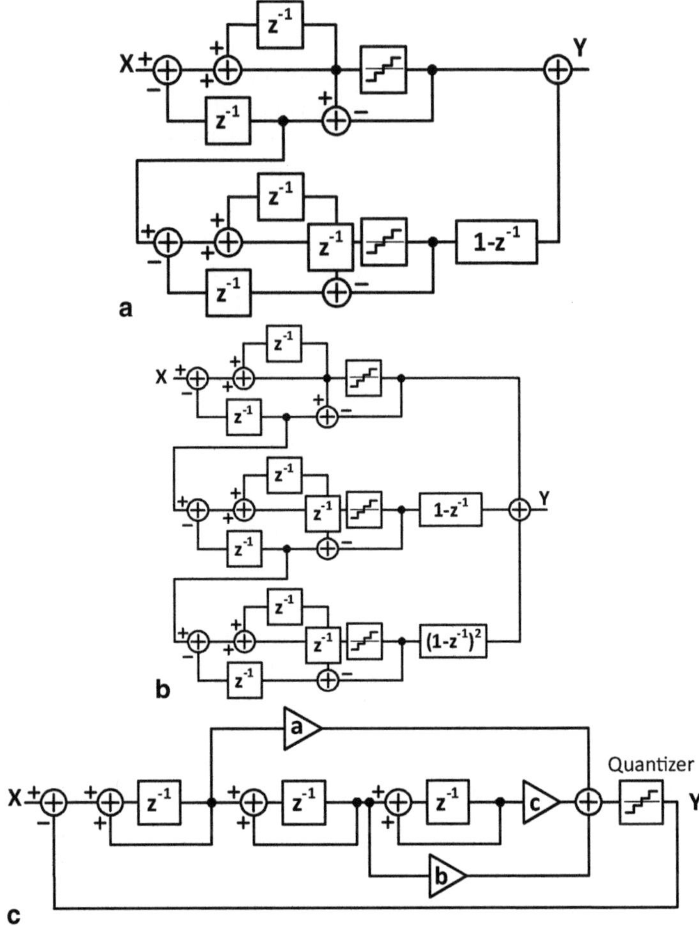

Fig. 2.2 $\Sigma\Delta$ modulator structures: **a** MASH1-1, **b** MASH1-1-1, and **c** SSMF

The output range of the SSMF structure is smaller than that that of the MASH1-1-1 structure. The PLL with SSMF structure also shows smaller instantaneous phase error at the input of PFD. The output noise spectrum should be compared to find the optimum $\Sigma\Delta$ modulator structure for a specific application. Figure 2.3 shows the output spectrum of the abovementioned four $\Sigma\Delta$ modulators. It is obvious that the third-order $\Sigma\Delta$ modulators provide better noise shaping effect than the second-order modulator. Among the third-order $\Sigma\Delta$ modulators, MASH1-1-1 structure has the minimum noise at low frequency offset but highest noise at high frequency offset. The choice of $\Sigma\Delta$ modulator topology depends on the specifications in its real applications. Ideally if better in-band noise is the goal, then MASH1-1-1 would be the best option. On the contrary, SSMF structure should be selected if the noise specification at half the sampling frequency is important. However, this intuitive choice is not always true when considering the nonlinearity of PLL loop, which will be discussed later.

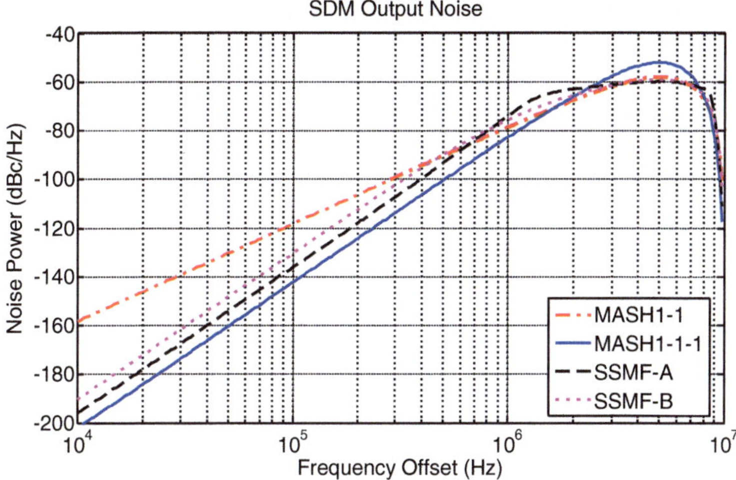

Fig. 2.3 Noise power of different ΣΔ modulators with 10 MHz sampling clock frequency

The quantization noise power shown in Fig. 2.3 is actually the shaped noise at the output of the ΣΔ modulators. To evaluate the impact of ΣΔ modulator noise, the quantization noise is converted to the phase noise spectrum at the PLL output. A ΣΔ modulator-based fractional-N PLL constantly dithers the divider value at a high rate compared to the bandwidth of the loop. An integrator is included before adding to the PLL loop as shown in Fig. 2.4 to convert the frequency domain noise to phase domain signals, whereas the control signal from ΣΔ modulator causes an instantaneous change in the frequency of the divider output.

By including the effect of integrator, the phase noise spectrum for MASH1-1-1 structure should be expressed as [11]

$$N^2_{\mathrm{MASH111}}(f) = \frac{\pi^2}{3}\left[2\sin\left(\pi\frac{f}{f_s} \right) \right]^{2(3-1)} \tag{2.3}$$

where white quantization noise spectra is assumed. Then, the phase noise contribution from quantization noise is low-pass filtered before reaching the PLL output.

Fig. 2.4 PLL model including ΣΔ modulator noise

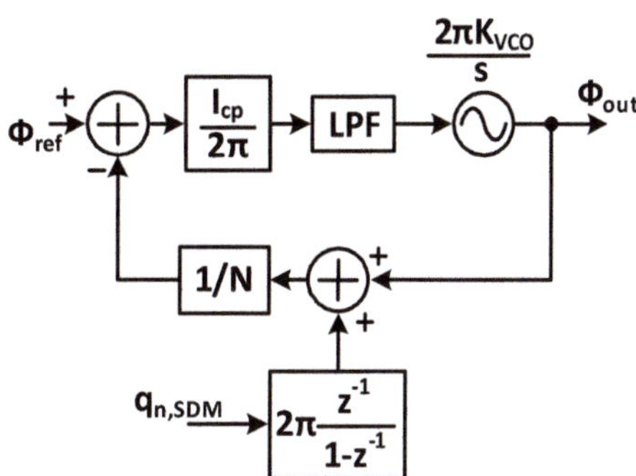

2.2.2   *Nonlinearity Analysis for ΣΔ Modulators*

The previous analysis is based on linear-time invariant model. The nonlinear effects of the PLL loop, especially gain mismatch of the charge pump caused by channel length modulation or dynamic switching, is not included in the model. However, the loop nonlinearity has very significant impact on the PLL phase noise caused by ΣΔ modulator. A phase-domain behavioral model as shown in Fig. 2.5 is used to analyze the nonlinear effect on ΣΔ modulator noise contribution.

Figure 2.6 shows the simulated phase noise spectrum of the four types of ΣΔ modulators with 3 % absolute gain mismatch in the nonlinear transfer function whereas the gain mismatch is defined as

$$\delta = \frac{I_{pos} - I_{neg}}{I_{pos} + I_{neg}} \tag{2.4}$$

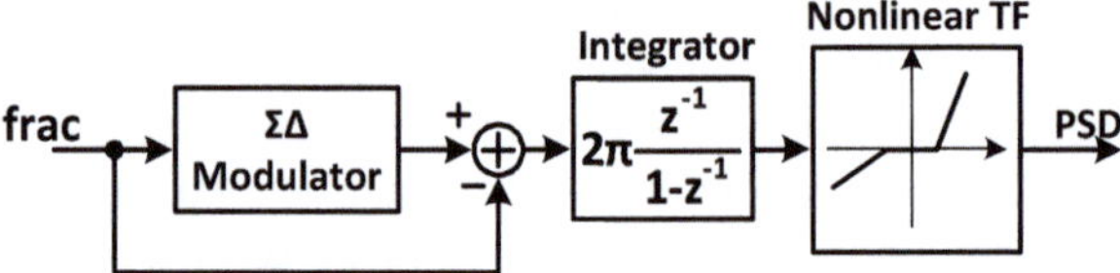

Fig. 2.5 Behavioral model to examine the nonlinearity effect on the ΣΔ modulator quantization noise

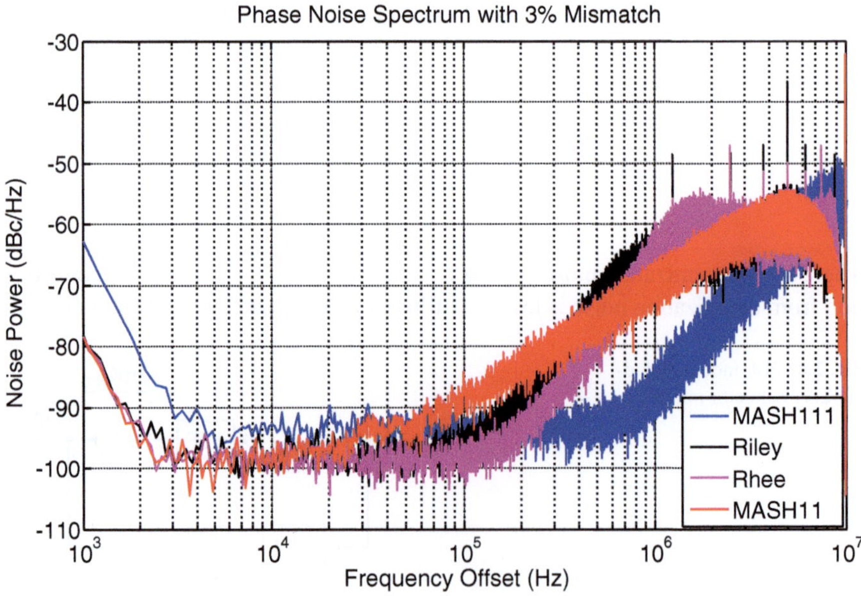

Fig. 2.6 Phase noise spectrum of MASH1-1 and MASH1-1-1 with 3 % gain mismatch in the transfer function

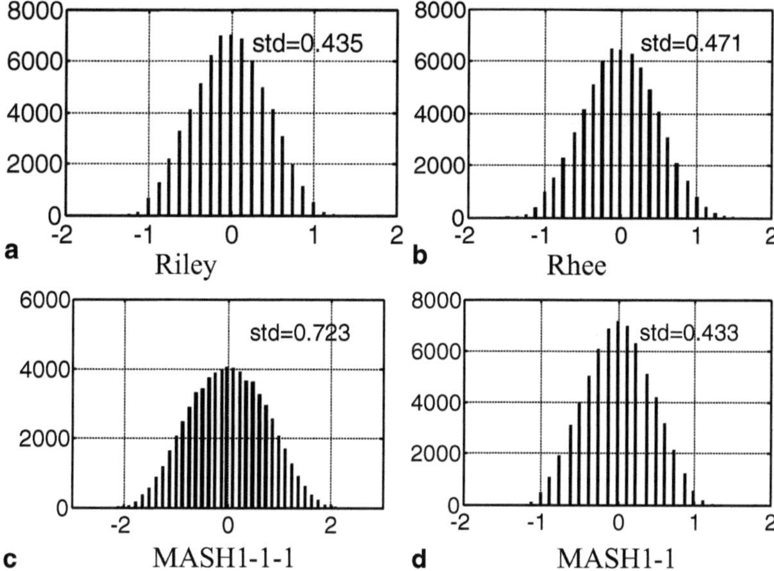

Fig. 2.7 Distribution of phase error at the input of PFD for different ΣΔ modulators

where I_{pos} and I_{neg} are the absolute positive gain and negative gain, respectively. From the simulation results, we can see that the in-band noise degradation for MASH1-1-1 modulator is higher than all the other three structures. This disadvantage is caused by its large output range from -3 to $+4$. The noise folding at low frequency offset for the two SSMF structures are very close to each other because their output range is similar. Both the two SSMF structure have fractional spurs above 1-MHz frequency offset while Riley's structure shows less spurious tones. Therefore, the choice of ΣΔ modulator structures is highly dependent on the noise specifications of the real application.

As a matter of fact, the in-band noise degradation can also be explained by observing the standard deviation of the ΣΔ modulator output range. Figure 2.7 shows the distribution and standard deviation of the four ΣΔ modulator structures. MASH1-1-1 structure exhibits an instantaneous phase error range of $-2\sim+2$ while the other three structures show an error range around $-1\sim+1$. The standard deviation for MASH1-1-1 is largest among the four structures, which explains why it has the highest in-band noise degradation due to the loop nonlinearity. The larger the output range, the worse the noise folding at low frequency offset. Therefore, it is highly desirable to improve the linearity of charge pump circuit since the noise folding is directly affected by its nonlinearity.

2.2.3 Quantization Noise Reduction Techniques

One of the simplest techniques used to reduce the nonlinearity is to improve the linearity of the charge pump circuit. The best achievable linearity is limited by the technology process and charge pump structure. A popular charge pump linearization

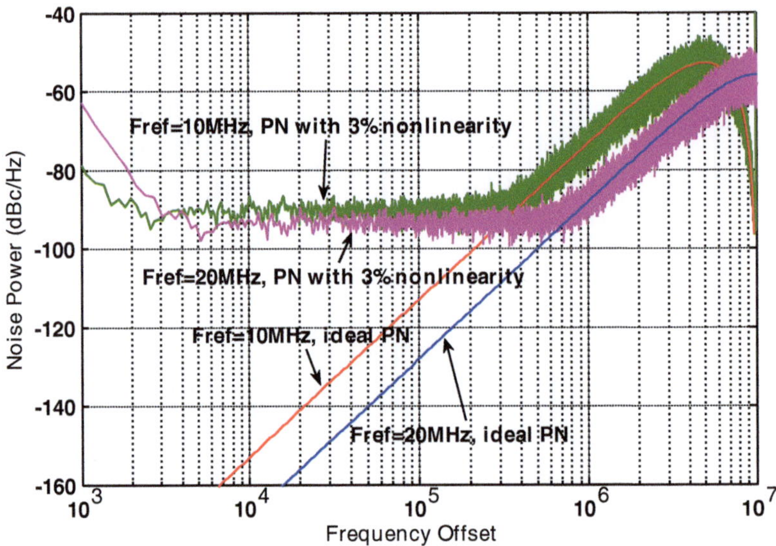

Fig. 2.8 Simulated noise improvements by doubling the clock frequency under 3 % gain mismatch

technique is to add a constant offset current into the charge pump and shift the transfer function [12]. Then, the PLL will allow the charge pump operating at only its positive or negative transfer function which has very good linearity performance.

Another simple but efficient technique for phase noise improvement is to double the reference frequency [13]. Ideally the phase noise can be improved by

$$\Delta_{\mathrm{PN}} = 6n - 3\mathrm{dB} \tag{2.5}$$

where n is the order of the $\Sigma\Delta$ modulator. At far-away frequency offset, the improvement follows Eq. (2.5). However, the noise improvement for the in-band noise is less than the expected value because of the nonlinearity. It has only 3 dB improvement at low frequency offset as shown in Fig. 2.8.

2.3 A Novel Noise Cancellation Technique for Fractional-N PLL

This section discusses a new quantization noise cancellation technique for high-order $\Sigma\Delta$modulator-based fractional-N PLL. Similar to conventional compensation technique with DAC currents, the proposed noise cancellation technique also injects current pulses into the loop filter. Usually the phase error can be compensated in three ways: (a) pulse-width modulated (PWM) current pulses; (b) pulse-amplitude modulated (PAM) current pulses; (c) combination of PWM and PAM.

The conventional fractional-N PLL can be modified by adding only a small cost of digital control logic to implement the proposed noise compensation technique. Typically a multi-modulus divider is used to divide the high-speed VCO signal to reference frequency. The divider in the fractional-N PLL will up or down count N-cycle of VCO period to implement the N-divider. This feature can be utilized to produce PWM current. With an auxiliary counter, a current pulse with X VCO cycles can be injected into the loop filter and compensate the instantaneous phase error. The main limitation for this technique is that X should be smaller than the division ratio of N. Moreover, the PLL in-band noise may increase because the long on-time of the DAC current injects more thermal noise into loop filter.

Another technique using PAM current pulses can also be implemented with current DAC. This technique is useful for fractional-N PLL with phase error in the range of $0 \sim 1$ VCO periods. However, it requires large area of current DAC circuit to compensate phase error in the range of $-2 \sim 2$ VCO periods for high-order $\Sigma\Delta$ modulators. Therefore, a compensating circuit that can generate both positive and negative phase error is preferred to cancel the quantization noise.

Figure 2.9a shows the system diagram of the proposed fractional-N PLL using the combination of PAM and PWM current pulses for high-order $\Sigma\Delta$ noise cancellation. The charge pump produces PAM signal while the PWM signal is produced by the pulse generation module. The PFD block, as shown in Fig. 2.9b, produces up and down control signal for the switches in the charge pump.

One detailed implementation of pulse control module is illustrated in Fig. 2.10. Only 2-bit pulse width control is used in this compensation scheme, other lower bits are implemented by DAC current. For example, if 6-bit DAC current is used, then the compensation algorithm can achieve an accuracy of 8-bit resolution by combining the pulse width control and DAC current injection. The area cost is relatively smaller than compensating techniques using only DACs.

Figure 2.11 illustrates the operating principles of the proposed noise compensation algorithm with phase error in the range of $(-2, 2)$ times T_{VCO}. The turn-on time of the down current equals to $T_X + 2T_{VCO}$ under locked condition. Take the second cycle for example, the phase error equals to $1 + \varepsilon T_{VCO}$, then integrated current in that comparison period can be expressed as

$$Q_{p2} = -I \cdot \left[T_X + (1 + \varepsilon)T_{VCO} - 4T_{VCO} + (1 - \varepsilon)T_{VCO} \right] + I \cdot t_d$$
$$= I \cdot (T_X - 2T_{VCO} + t_d) \tag{2.6}$$

Similarly, the integrated current in the fourth comparison period can be expressed as

$$Q_{p4} = -I \cdot \left[T_X - (1 + \varepsilon)T_{VCO} - T_{VCO} + (1 - \alpha)T_{VCO} \right] + I \cdot t_d$$
$$= I \cdot (T_X - T_{VCO} + t_d - \varepsilon T_{VCO} - \alpha T_{VCO}) \tag{2.7}$$

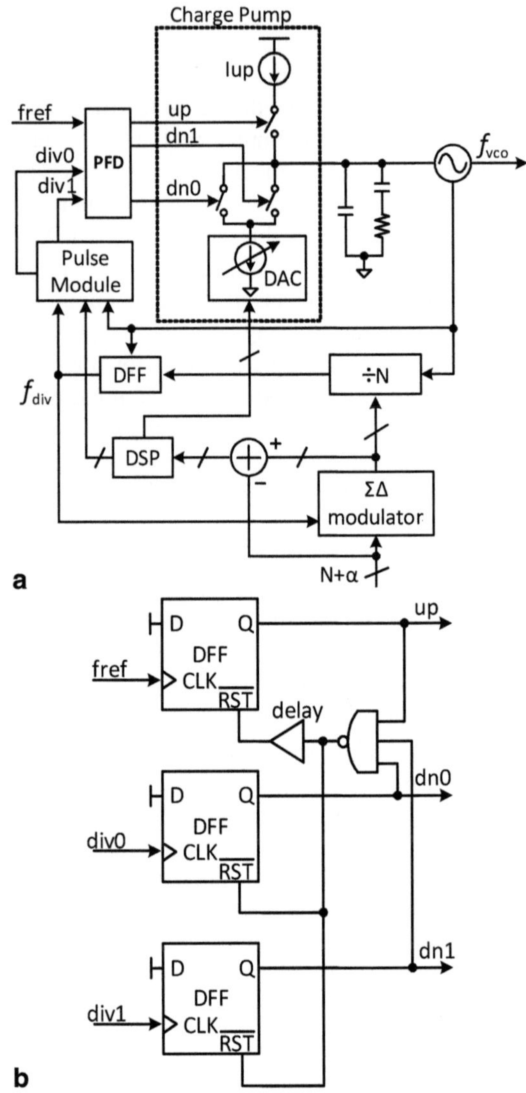

Fig. 2.9 **a** System diagram of the proposed fractional-N PLL with quantization noise cancellation technique; **b** PFD circuit

To completely remove the quantization noise caused by the $\Sigma\Delta$ modulator, the integrated charge should be constant and equal to zero. Therefore, we can arrive at the following equations:

$$\alpha = 1 - \varepsilon \tag{2.8}$$

$$T_X - 2T_{VCO} + t_d = 0 \tag{2.9}$$

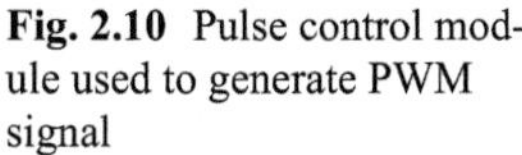

Fig. 2.10 Pulse control module used to generate PWM signal

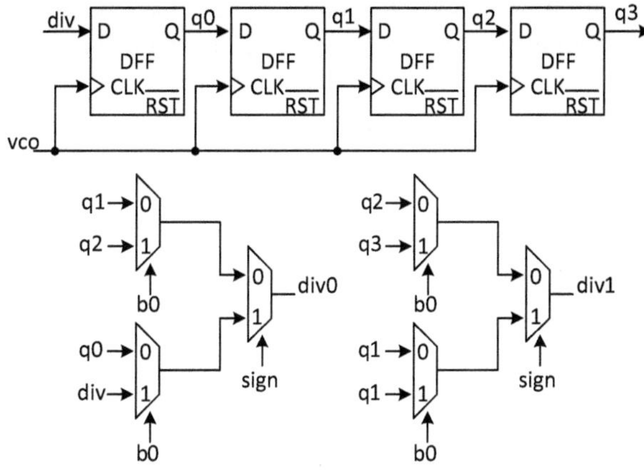

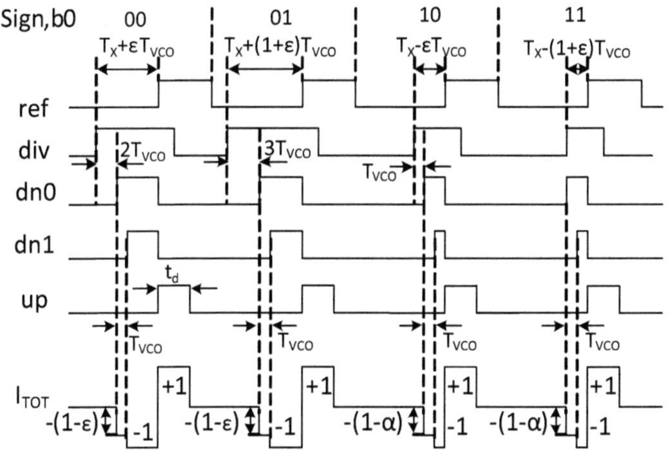

Fig. 2.11 Waveform example for phase error compensation

where α is defined in Fig. 2.11, T_X is the intrinsic delay in the PLL loop, and t_d is the delay introduced in the reset path of up control flip-flop. Typically, we can choose t_d slightly larger than twice the VCO period to allow correct compensation. With the proposed noise cancellation technique, the noise contribution from $\Sigma\Delta$ modulator can be eliminated under ideal condition.

2.4   Conclusion

This chapter reviewed several quantization noise reduction techniques for $\Sigma\Delta$ modulator-based fractional-N PLL. Analysis of quantization noise and nonlinearity effect is given for MASH1-1, MASH1-1-1, and two SSMF $\Sigma\Delta$ modulators. Several quantization noise reduction techniques have been discussed and it shows that the most efficient way of reducing $\Sigma\Delta$ quantization noise is frequency doubling. A novel noise cancelling technique and its implementation for second-and third-order $\Sigma\Delta$ modulator has been proposed to compensate the phase error between $-2\sim2$ VCO period.

References

1. T. A. D. Riley, M. A. Copeland, T. A. Kwasniewski, "Delta-sigma modulation in fractional-N frequency synthesis," IEEE J. Solid-State Circuits, vol. 28, no. 5, pp. 553–559, May. 1993.
2. W. Rhee, B. Song, A. Ali, "A 1.1-GHz CMOS fractional-N frequency synthesizer with a 3-b third-order $\Delta\Sigma$ modulator," IEEE J. Solid-State Circuits, vol. 35, no. 10, pp. 1453–1460, Oct. 2000.
3. B. D. Muer and M. S. J. Steyaert, "A CMOS Monolithic $\Delta\Sigma$-controlled fractional-N frequency synthesizer for DCS-1800," IEEE J. Solid-State Circuits, vol. 37, no. 7, pp. 835–844, Jul. 2002.
4. P. Su and S. Pamarti, "Mismatch shaping techniques to linearize charge pump errors in fractional-N PLLs," IEEE Transactions on Circuits and Systems-I: Regular papers, vol. 57, No. 6, pp. 1221–1230, Jun. 2010.
5. T. Lin, C. Ti, Y. Liu, "Dynamic current-matching charge pump and gated-offset linearization technique for delta-sigma fractional-N PLLs," IEEE Transactions on Circuits and Systems-I: Regular papers, vol. 56, No. 5, pp. 877–885, May. 2009.
6. A. Swaminathan, K. J. Wang, and I. Galton, "A wide-bandwidth 2.4 GHz ISM band fractional-N PLL with adaptive phase noise cancellation," IEEE J. Solid-State Circuits, vol. 42, no. 12, pp. 2639–2649, Dec. 2007.
7. S. Pamarti, L. Jansson, and I. Galton, "A Wideband 2.4-GHz Delta-Sigma Fractional-N PLL With 1-Mb/s In-Loop Modulation," IEEE J. Solid-State Circuits, vol. 39, No. 1, pp. 49–62, Jan. 2004.
8. S. E. Meninger and M. H. Perrott, "A 1-MHZ Bandwidth 3.6-GHz 0.18-um CMOS Fractional-N Synthesizer Utilizing a Hybrid PFD/DAC Structure for Reduced Broadband Phase Noise," IEEE J. Solid-State Circuits, vol. 41, pp. 966–980, Apr. 2006.
9. M. H. Perrott, T. L. Tewksbury III, C. G. Sodini, "A 27-mW CMOS fractional-N synthesizer using digital compensation for 2.5-Mb/s GFSK modulation," IEEE J. Solid-State Circuits, vol. 32, no. 12, pp. 2048–2060, Dec. 1997.
10. S. B. Sleiman, J. G. Atallah, S. Rodriguez, A. Rusu, M. Ismail, "Optimal $\Sigma\Delta$ modulator architectures for fractional-N frequency synthesis," IEEE Transactions on VLSI Systems, vol. 18, no. 2, pp. 194–200, Feb. 2010.
11. M. H. Perrott, M. D. Trott, C. G. Sodini, "A modeling approach for Σ-Δ fractional-N frequency synthesizers allowing straightforward noise analysis," *IEEE J. Solid-State Circuits*, vol. 37, no. 8, pp. 1028–1038, Aug. 2002.
12. H. Huh, Y. Koo, K. Lee, Y. OK, S. Lee, D. Kwon, J. Lee, J. Park, K. Lee, D. Jeong, W. Kim, "Comparison frequency doubling and charge pump matching techniques for dual-band $\Delta\Sigma$ fractional-N frequency synthesizer," *IEEE J. Solid-State Circuits*, vol. 40, no. 11, pp. 2228–2236, Nov. 2005.
13. W. Lee and S. Cho, "A 2.4-GHz reference doubled fractional-N PLL with dual phase detector in 0.13-μm CMOS," *Proceedings of Circuits and Systems (ISCAS)*, pp. 1328–1331, 2010.

Chapter 3
A Wide-Band Low Power BiCMOS PLL

3.1 Background

With the development and improvements of semiconductor process and technology, low--cost consumer applications for radar transceiver systems have been proposed, such as automotive collision avoidance, cruise control, and fluid-level indicators. High performance local oscillator (LO) signals with clean spectrum are required to down-convert the received radio-frequency (RF) signal and up-convert the low frequency base-band signal to RF signal for radar transceivers. Phase-locked loop (PLL) is such a system that can be used to generate LO signals with large frequency range by tuning the integer or fractional divider. This chapter presents an integer-N PLL system comprised of a wide-band VCO, divide-by-two circuit (DTC), multi-modulus divider (MMD), phase-frequency detector (PFD) and charge pump (CP) in current-mode logic (CML) type, and divider modules for mixers and direct digital synthesis (DDS) clock.

Among these building blocks of PLL system, VCO plays a very important role in generating a clean spectrum since it dominates the out-of-band phase noise performance. Recently, a tail current shaping technique has been proposed to improve the phase noise of conventional CMOS VCOs [1]. Similarly, the current shaping technique can also be used to improve the noise performance of bipolar VCOs. With an optimized filtering capacitor at the current tail, the VCO phase noise can be improved by 3.5 dB according to the Spectre simulation results. Usually, the smaller the VCO tuning gain, the better the phase noise performance of the VCO. In order to improve the phase noise performance and extend frequency tuning range, the overall frequency range can be divided into 16 or more sub-bands controlled by MOS switches [2]. However the utilization of MOS switches will increase the parasitic capacitance to the VCO load and thus narrow the frequency tuning range. To address the problem, static bias voltage is applied to those parasitic diodes.

Frequency divider is another critical building block for high-performance clock synthesizers. In order to divide the high-frequency VCO output signal, the divider speed and power trade-off should be carefully considered at an early design phase. Minimum power consumption for CMOS dividers based on CML can be achieved

© Springer International Publishing Switzerland 2015

F. Zhao, F. F. Dai, *Low-Noise Low-Power Design for Phase-Locked Loops,*
DOI 10.1007/978-3-319-12200-7_3

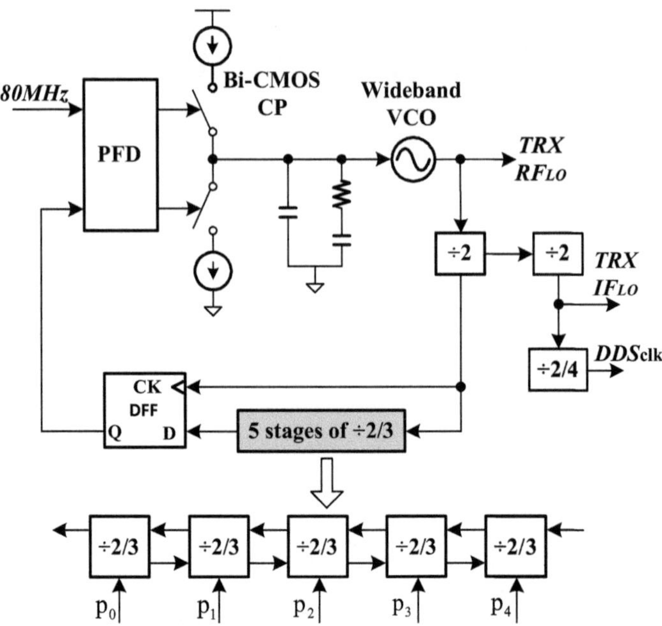

Fig. 3.1 System diagram of the wideband PLL system

by finding out the optimum transistor size and operating point [3] through compli-
cate simulations. However, it is not intuitive to use such an optimization methodology for divider design in real applications.

Typically a divider circuit has a self-oscillation frequency F_{osc} which requires
minimum input power to divide the input signal [4]. The self-oscillation frequency
is better to be equal to the target input frequency since this requires minimum power
consumption to achieve proper divider function. To find the F_{osc}, the frequency at
which the DTC and divide-by-2/3 cell resonate, a simplified design technique based
on timing delay analysis is developed. With the proposed optimization technique, a
CML divider chain can achieve minimum power consumption in either CMOS or
bipolar technology. After the minimum input power at F_{osc} of dividers is found, the
signal swing should be kept large enough to cover the process, voltage, and temperature (PVT) variations. On the other hand, the input signal swing should be kept
relatively large for better output phase noise performance. To achieve the optimum
divider design, it is desirable to understand the tradeoff between the phase noise,
power and speed.

The proposed PLL architecture is shown in Fig. 3.1. The output signal of the
VCO is first divided by 2 before feeding to the MMD to relax the timing requirement. Five stages of divide-by-2/3 cell similar to [5] are used to achieve a divide
ratio ranging from 32 to 63. The LO signal for the first RF mixer is connected to
the buffer loading the VCO. One additional DTC is adopted to generate the second

LO signal for IF mixer and DDS block. Different from conventional PLL structures which use CML-to-CMOS level converter [6], the divided VCO signal is directly fed to the CML type phase detector to reduce the reference spur. The tri-state PFD and charge pump is also implemented in CML with a signal swing of 200 mV. Moreover, in order to reduce the reference spur caused by the mismatch of sinking and sourcing current and to increase the voltage headroom, a Bi-CMOS charge pump is adopted.

The organization of this chapter is as follows. Detailed analysis for bandgap voltage reference design is given in Sect. 3.2. In Sect. 3.3, the design of the wideband VCO and divider are discussed in depth. Moreover, the proposed optimization methodology of DTC, divide-by-2/3 cell, and whole divider chain are explored in detail. Then, the circuit implementation and measurement results are described in Sect. 3.4. Finally, conclusions are drawn in Sect. 3.5.

3.2 Analysis of Bandgap Reference for Current Generation

Fully integrated bandgap circuits with high power supply rejection ratio (PSRR) are indispensable to provide clean reference voltage and current for RF front-end modules. In order to guarantee reliable start-up condition, classic all NPN bandgap reference circuit is adopted since it does not require operational amplifier (OPAMP) and start-up circuit [7]. Figure 3.2 shows the bandgap circuit for current reference generation. Different from the classic structure, a NMOS transistor with small threshold voltage is used to enhance the PSRR. The voltage at node A shown in Fig. 3.2 is

$$V_{A} = V_{BE1} + (I_{B1} + I_{C2})R_{2} = V_{BE3} + (I_{B2} + I_{B3} + I_{C3})R_{3}. \tag{3.1}$$

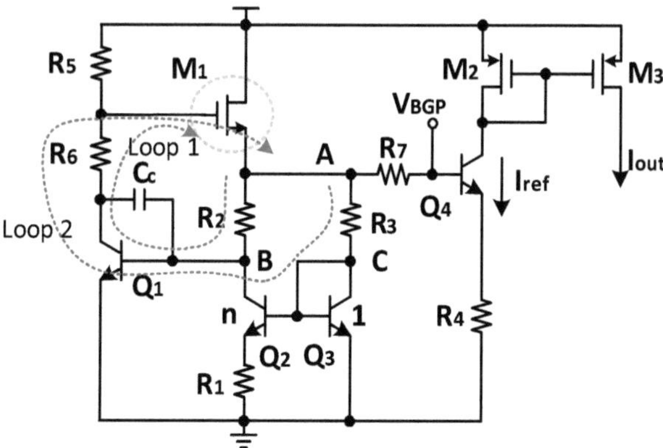

Fig. 3.2 Bandgap circuit utilized to generate voltage reference and current reference

The voltage difference of V_{BE2} and V_{BE3} produce a current following through resistor R_1 expressed as

$$I_{C2} = \frac{V_{BE3} - V_{BE2}}{R_1} = \frac{V_T}{R_1}\left(ln\frac{I_{C3}}{I_{S3}} - ln\frac{I_{C2}}{I_{S2}}\right) = \frac{V_T}{R_1}ln\left(n\frac{I_{C3}}{I_{C2}}\right). \tag{3.2}$$

Suppose the current gain β is very large and Q_1, Q_3 have the same size, the voltage at base of Q_1 and Q_2 will be similar to each other because of the negative feedback loop formed by Q_1, R_5, and M_1. With $V_{BE1} = V_{BE3}$, we can obtain the following equation

$$I_{C2}R_2 \approx I_{C3}R_3 \Rightarrow \frac{I_{C2}}{I_{C3}} = \frac{R_3}{R_2}. \tag{3.3}$$

Then, the bandgap voltage at node A can be written as

$$V_{BGP} = V_A = V_{BE3} + V_T\frac{R_2}{R_1}ln\left(n\frac{R_2}{R_3}\right). \tag{3.4}$$

Note that there is a positive feedback loop 2 in the circuit. To stabilize the circuit, the positive loop gain should be smaller than the negative loop gain. Assuming that β is very large, we have the following condition for stable operation:

$$\frac{1}{1+g_{m,Q3}R_3}\frac{g_{m,Q2}R_2 + g_{m,Q2}\,/\,g_{m,M1}}{1+g_{m,Q2}R_2} < 1. \tag{3.5}$$

The PSRR under dc condition can be derived and approximated as

$$PSRR_{DC} \approx 1 + g_{m,Q1}R_5 + \frac{g_{m,Q3}}{g_{m,M1}(1+g_{m,Q3}R_3)}. \tag{3.6}$$

In order to improve the PSRR, we need to increase the resistor value of R_5; however, the voltage drop on R_5 is limited by the supply voltage. Therefore, a NMOS transistor, whose V_{GS} voltage is smaller than the V_{BE} of NPN transistor, is used to increase the PSRR performance.

It is worth mention that the current following through M_1 is the sum of the current through Q_2 and Q_3, so we can further simply the PSRR equation to be

$$PSRR_{DC} \approx 1 + \frac{I_{C1}}{V_T}R_5 + \frac{\dfrac{I_{C3}}{V_T}}{\dfrac{2(I_{C2}+I_{C3})}{V_{OD1}}\left(1 + \dfrac{I_{C3}}{V_T}R_3\right)}$$

$$\approx 1 + \frac{V_{DD}-V_{BGP}-V_{GS1}}{V_T} + \frac{0.5 V_{OD1}}{(1+\alpha)(V_{BGP}-V_{BE3}+V_T)}, \tag{3.7}$$

where α is the ratio of I_{C2} to I_{C3}, V_{OD1} is the overdrive voltage of M_1. From the above equation, we can see that the PSRR of this architecture is mainly limited by supply voltage. The PSRR can be improved by choosing relatively small V_{GS1}. That is the main reason to use a NMOS transistor to replace the NPN transistor in the classic bandgap circuit. It seems that larger V_{BE3} or smaller α can also lead to better PSRR from the third term of Eq. (3.7). However, the third term is relatively small compared with the second item, for instance, with $V_{BGP}=1.2\,V$, $V_{BE}=0.8\,V$, $\alpha=0$, and $V_{OD1}=0.4\,V$, the third item is 0.24. Thus, the third item has very little impact on the overall PSRR.

3.3 Circuit Design of Critical Blocks

3.3.1 VCO with Extended Frequency Range

Figure 3.3 shows the schematic of the proposed bipolar NPN cross-coupled VCO. To achieve both wide frequency tuning range and low phase noise performance, a 16-subband VCO with 4-bit MIM capacitor array and tunable current tail is utilized. To improve the phase noise performance, a noise filtering MIM capacitor C_{filt} is added in parallel with the current tail to reduce the noise up-conversion from the NMOS current tail and the simulated improvement is around 3.5 dB.

A symmetric inductor with deep trench is adopted for better isolation from substrate noise coupling. Conventional switches used for coarse tuning between different VCO bands usually consist of three NMOS transistors [2]. However, the parasitic capacitances resulted from parasitic drain (n+)-to-substrate (p+) diodes vary with the dc voltage applied to its drain, i.e., the smaller the dc bias voltage, the larger the parasitic capacitance. When the MIM capacitors are turned off, i.e., NMOS switching transistors are in off state, the drain of the NMOS switch would be floating with zero DC voltage and thus large parasitic capacitance loads the LC tank. In order to decrease the series resistor and increase the quality factor of the capacitor array, the NMOS transistors for switches are usually implemented with large dimensions. Then, the large tuning range is suffered from these parasitic drain-substrate diodes. In order to widen the VCO frequency tuning range, a dc voltage equal to supply voltage is applied to the drain through small PMOS transistors as

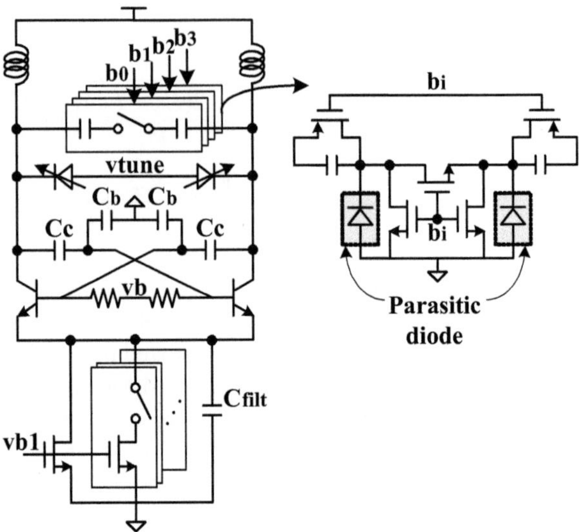

Fig. 3.3 Schematic of the proposed VCO with extended tuning range

shown in Fig. 3.3. When the NMOS switches are off, a high dc bias voltage is applied to the drain of NMOS transistors. With the proposed frequency extending technique, the simulated frequency tuning range can be increased from 2.32 to 2.74 GHz, which is 20 % improvement.

The bipolar transistor can easily break down with a 2.2 V supply voltage since its collector–emitter break down voltage V_{ceo} is 1.6 V. In order to avoid transistor break down, the bias voltage vb as shown in Fig. 3.3 should be large. But large vb will force the bipolar transistor into saturation region which is unwanted since the VCO phase noise will be degraded in this operating region. The following equations should be met to maintain the largest signal swing for best phase noise performance while operating in safe region:

$$\begin{cases} vb < V_{\text{DD}} - (1+\text{n})A_{\text{VCO}} - V_{\text{margin}} & \text{saturation} \\ vb > V_{\text{DD}} + (1+\text{n})A_{\text{VCO}} + V_{\text{th}} - V_{\text{ceo}} & \text{break down}, \end{cases} \tag{3.8}$$

where ratio $n = C_C / (C_C + C_b)$ is introduced to alleviate the requirement of break-down voltage. A_{VCO} is the single-ended VCO signal amplitude. $V_{\text{margin}} = 0.2\,\text{V}$ and $V_{\text{th}} = 0.5\,\text{V}$ are the voltage margin from saturation and threshold voltage of bipolar transistor, respectively. According to the noise Eq. (66) of bipolar cross-coupled differential LC-tank VCO in reference [8], we know that the larger the voltage divider ratio n, the better the phase noise. To prevent break down, $n = 2/3$ is chosen and the corresponding maximum differential signal amplitude A_{VCO} is chosen to be 0.54 V for this implementation.

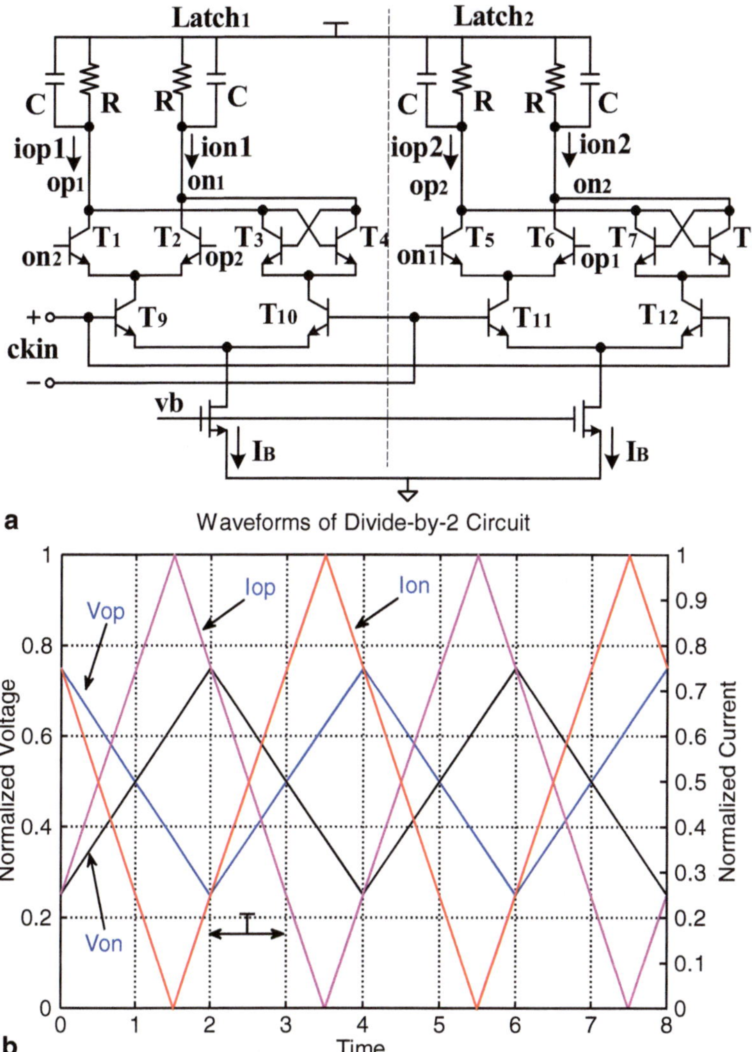

Fig. 3.4 Divide-by-2 circuit: **a** Circuit schematic and **b** simplified waveforms for the derivation of self-oscillation frequency

3.3.2 Optimization Methodology for DTC

Classic CML structure is used to implement the DTC and divide-by-2/3 cell. Figure 3.4 shows the DTC schematic and output waveforms under self-oscillation mode. Analytic design equation of F_{osc} is developed to find the optimum operating point for minimum power consumption and noise performance. The outputs of latch$_1$ and latch$_2$ have 90° phase difference and the output voltages can only swing from $V_{DD} - 0.25V_{SW}$ and $VDD - 0.75V_{SW}$. As shown in Fig. 3.4b, the corresponding current flowing through the load resistor and capacitor are $0.25I_B$ and $0.75I_B$ when V_{OP} equals to $V_{DD} - 0.25V_{SW}$ and $V_{DD} - 0.5V_{SW}$, respectively.

By assuming that $V_{OP} = V_{DD} - 0.25V_{SW}$, $V_{ON} = V_{DD} - 0.75V_{SW}$ at initial time $t=0$, the output voltage with time can be simplified to be

$$\begin{cases} V_{OP}(t) = V_{DD} - \dfrac{0.5V_{SW}}{T}[RC(e^{-t/RC} - 1) + t + 0.5T] \\ V_{ON}(t) = V_{DD} + \dfrac{0.5V_{SW}}{T}[RC(e^{-t/RC} - 1) + t + 1.5T] \, , \end{cases} \tag{3.9}$$

where the signal swing is $V_{SW} = I_B R$. By equating the above two equations and let $t=T$, we can get

$$e^{-\frac{T}{RC}} = 1 - \frac{0.5T}{RC}. \tag{3.10}$$

The above function can be solved with numerical method and the result is $T = 1.59RC$. Therefore, the self-oscillation frequency of the DTC can be expressed as

$$f_{osc,\,dtc} = \frac{1}{4T} = \frac{1}{4 \times 1.59RC} = \frac{0.157\text{GHz}}{R(\text{k}\Omega)C(\text{pF})}. \tag{3.11}$$

With Eq. (3.11), it is nontrivial to derive the load resistor when we know the estimated parasitic capacitance. Figure 3.5 illustrates the calculation and simulation results of F_{osc} for a DTC example. The simplified model predicts the self-oscillation frequency with an error less than 5%.

To ensure correct functionality for a DTC block, the gain of each latch stage should be large enough to sample and hold the input signal and the following condition needs to be met

$$\text{Gain}_{bip} = g_m R = \frac{0.5I_B}{V_T} \times R = \frac{0.5V_{SW}}{V_T} > 1. \tag{3.12}$$

Equation (3.12) can easily be met for bipolar dividers since the chosen signal swing V_{SW} is usually at least four times the thermal voltage $V_T = 26$ mV. Only minor modifications need to be made to Eq. (3.12) for CMOS divider circuits and it can be written as

$$\text{gain}_{cmos} = \frac{2 \times (0.5I_B)}{V_{GS} - V_{TH}} \times R = \frac{I_B \times R}{V_{GS} - V_{TH}} = \frac{V_{SW}}{V_{GS} - V_{TH}} > 1, \tag{3.13}$$

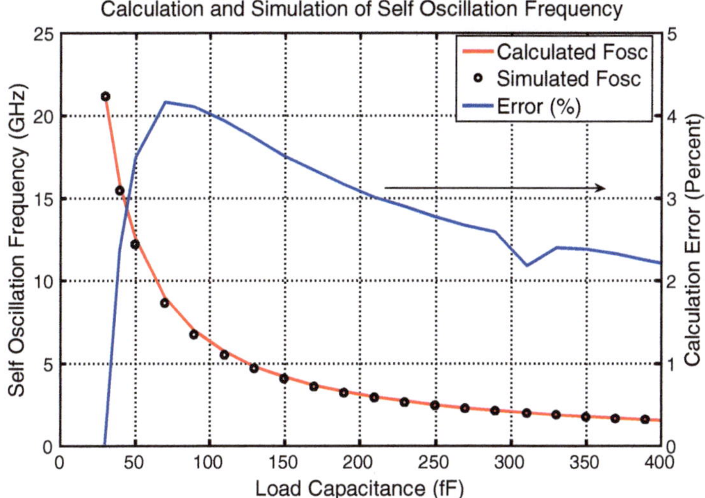

Fig. 3.5 Comparison of calculated and simulated self-oscillation frequency of DTC with $R = 250\,\Omega$ and $I_B = 0.8$ mA

where the g_m is a first-order approximation of MOS trans-conductance [9]. Similar to bipolar transistors, Eq. (3.13) can also be met automatically because the desired V_{SW} is usually larger than the overdrive voltage $V_{GS} - V_{TH}$ to fully switch the differential pairs.

Figure 3.6a shows the sensitivity curves of a DTC at three different bias condition and phase noise variation with input signal. The simulated F_{osc} is around $9.3/2 = 4.65$ GHz and the calculation result is 4.83 GHz which is 3.9 % higher than that obtained from simulation. On the other hand, the input signal should be large enough to avoid phase noise degradation since the phase noise performance of a DTC also depends on the input power as shown in Fig. 3.6b.

3.3.3   Design of Divide-by-2/3 and MMD

The other main building block for MMD is the divide-by-2/3 cell as shown in Fig. 3.7, which is made up of two groups of components: (1) gate G_1, latches DL_1 and DL_2 for divide-by-2/3; (2) gate G_2 and G_3, latches DL_3 and DL_4 for divide-by-3 mode only. The design of latch DL_1 and DL_2 can use the previous optimization method of DTC by considering the delay introduced by gate G_1. However, the design of divide-by-3 mode is different from DTC and it can be simplified by treating all the latches and gates as the same unit delay stages.

The whole divider chain for PLL includes one DTC and five stages of divide-by-2/3 cells. The most critical stage is the first DTC stage since it should be able to operate at the VCO frequency. With the proposed power optimization for DTC, we

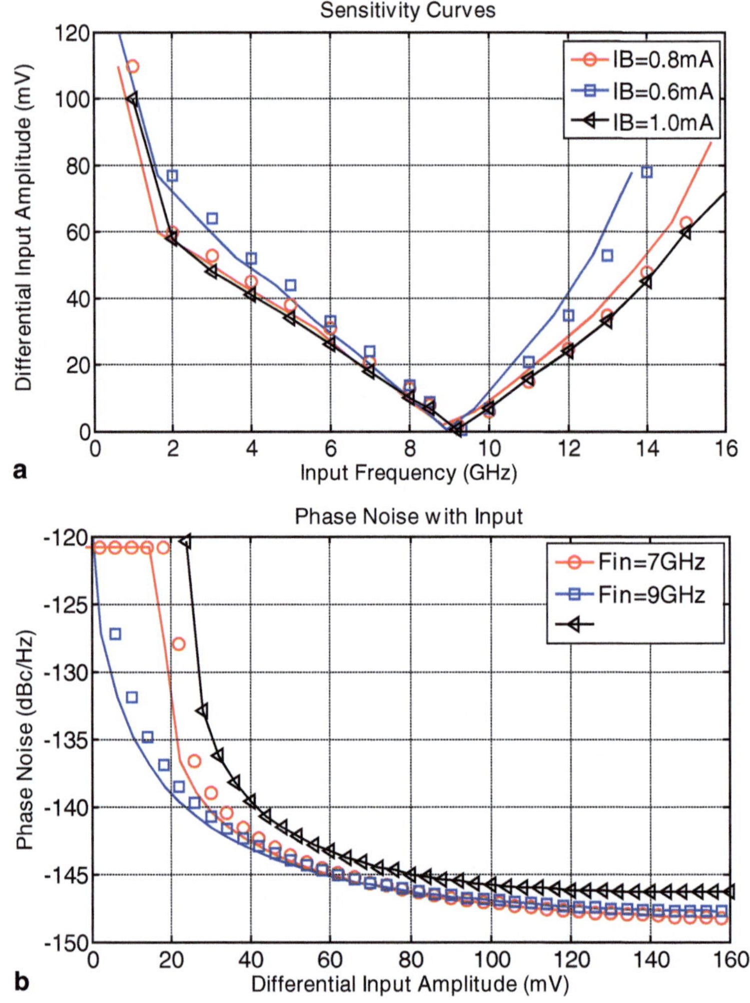

Fig. 3.6 Simulation results for DTC with $R=250\ \Omega$, $C=130$ fF: **a** Simulated sensitivity curves, and **b** phase noise performance with different input signal

can easily calculate the loading resistor value and power for the required speed. The power consumption for divide-by-2/3 cell for MMD can be scaled down with the input frequency. The output signal of the MMD is resynchronized with VCO divide-by-2 signal to eliminate the noise accumulated when going through the cascaded divider stages. Figure 3.8 shows the simulated waveforms at different output of the MMD block under low power mode which requires only 100–150 mV single-ended signal swing.

Fig. 3.7 Circuit schematic of divide-by-2/3 cell

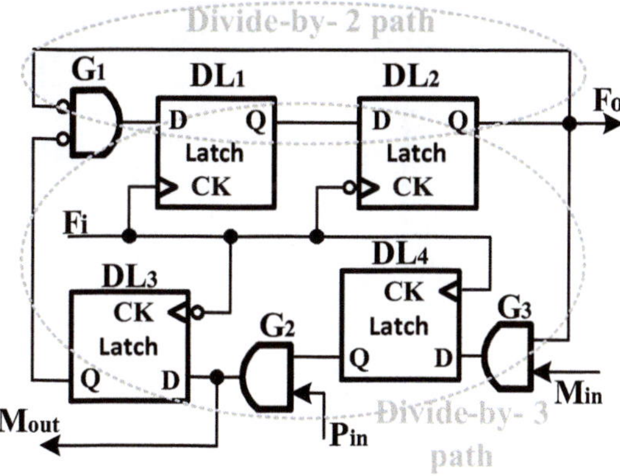

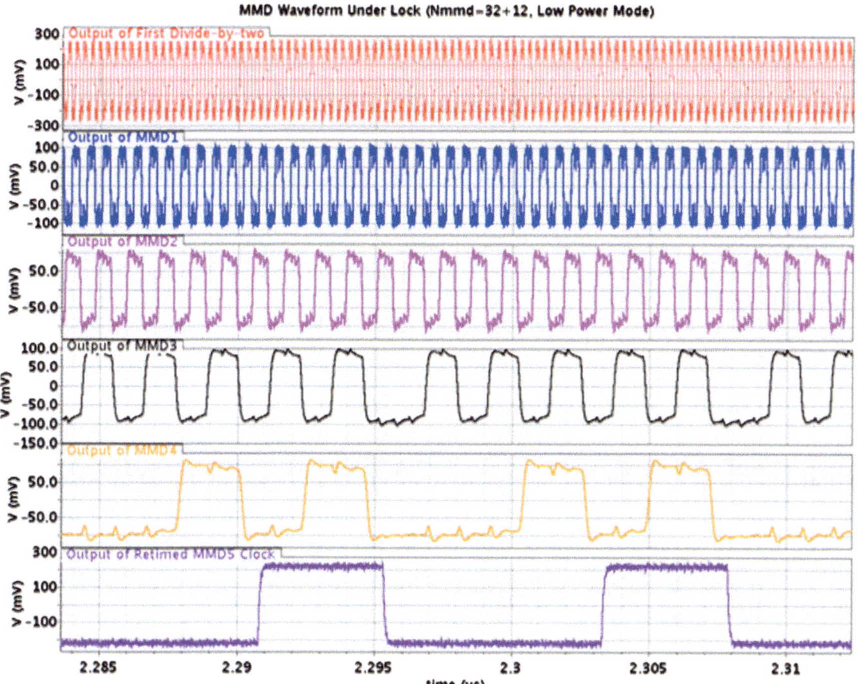

Fig. 3.8 Postlayout simulation result of MMD output signals

3.4   Experimental Results

The wide-band PLL was implemented in a 0.13 μm SiGe BiCMOS technology and the die photo of the core PLL is shown in Fig. 3.9. The PLL including driving buffers for T/Rx mixer and DDS clock occupies an area of 1.3×0.65 mm^2. The whole PLL consumes 32 mA from a 2.0–2.3 V supply voltage.

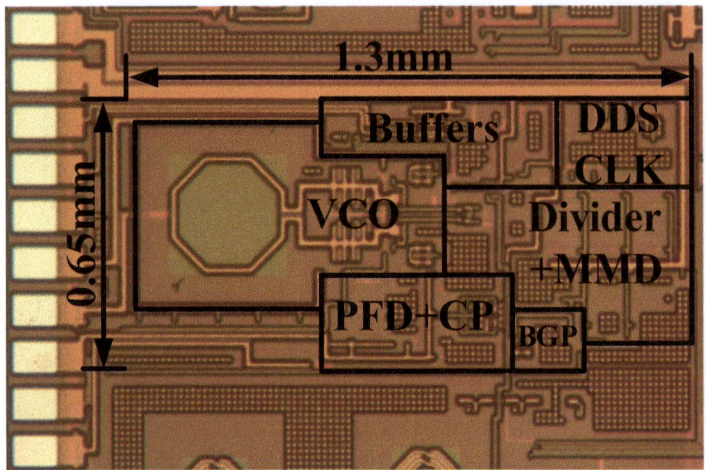

Fig. 3.9 Die photo of the implemented PLL

3.4.1   Phase Noise and Frequency Tuning Range

As shown in Fig. 3.10, the measured phase noise of the PLL is -86 and -114 dBc/ Hz @ 10 kHz and 1 MHz offset with a center frequency of 6.56 GHz, respectively. The measured VCO frequency tuning range is illustrated in Fig. 3.11. The overall frequency tuning range is around 34 % and the measured VCO tuning gain is around 350 MHz/V.

3.4.2   Output Spectrum and Lock Time

Shown in Fig. 3.12 is the spectrum measured at the PLL output. From the spectrum, it can be seen that the reference spur at 80 MHz offset is less than -65 dBc/Hz. This low-level reference spur is achieved by the reduced signal swing in the CML PFD and charge pump. The measured lock time for this PLL with a 100-kHz loop bandwidth is 15 µs as shown in Fig. 3.13.

3.5   Conclusion

A 4.8–6.8 GHz PLL with a power optimized multi-modulus divider (MMD) for radar applications is presented in this paper. Based on the timing delay analysis and the self-oscillation frequency of the divider cells, the power consumption of the di- vide-by-two circuit (DTC) and divide-by-2/3 cells can be minimized. To extend the frequency tuning range of the voltage controlled oscillator (VCO), PMOS switches

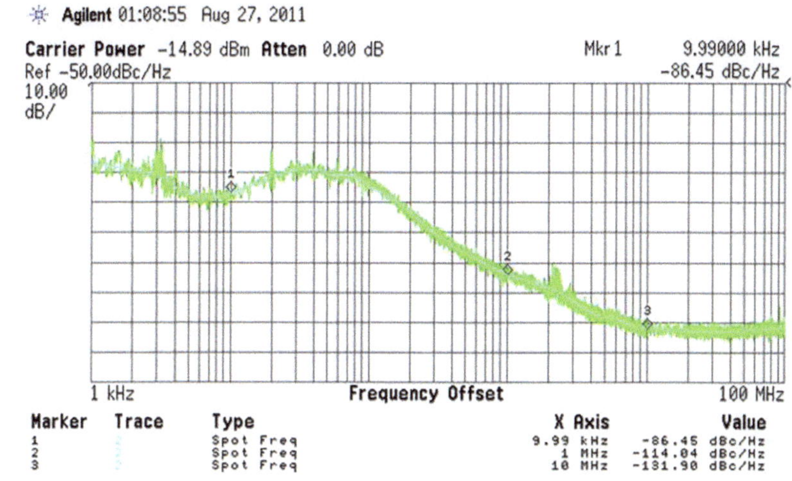

Fig. 3.10 Measured phase noise of the PLL with BW $=100$ kHz, $F_{ref}=80$ MHz

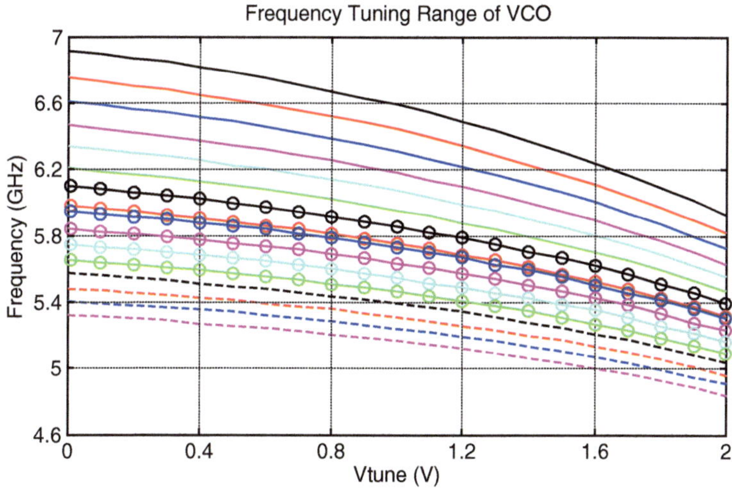

Fig. 3.11 Measured PLL frequency range

are used to reverse bias the parasitic diode. The proposed PLL achieves a measured phase noise of -86 dBc/Hz @10-kHz offset and -114 dBc/Hz@1-MHz offset with a center frequency of 6.56 GHz, and it consumes 64 mW from a 2.0 V supply voltage. The PLL system for the radar transceiver is implemented in 0.13 μm SiGe technology with a core area of 1.35×0.65 mm^2. The performance summary for the PLL frequency synthesizer is given in Table 3.1.

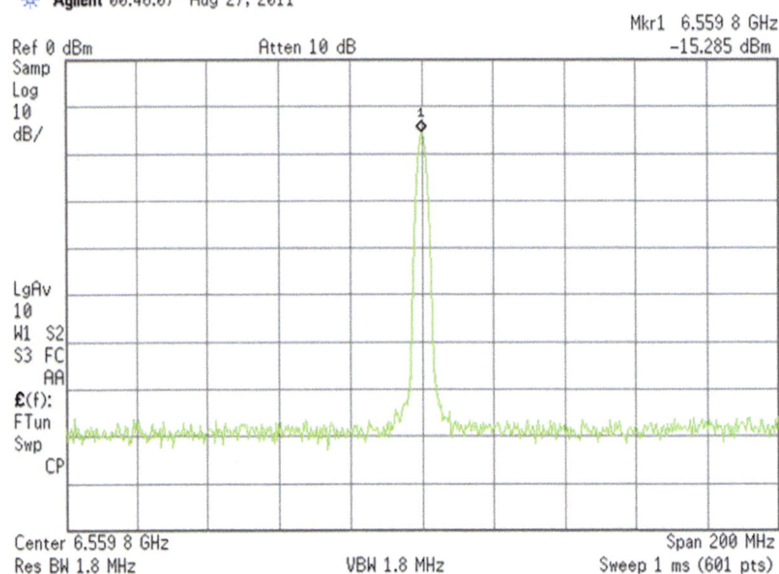

Fig. 3.12 PLL output spectrum

Fig. 3.13 Measured PLL locking time

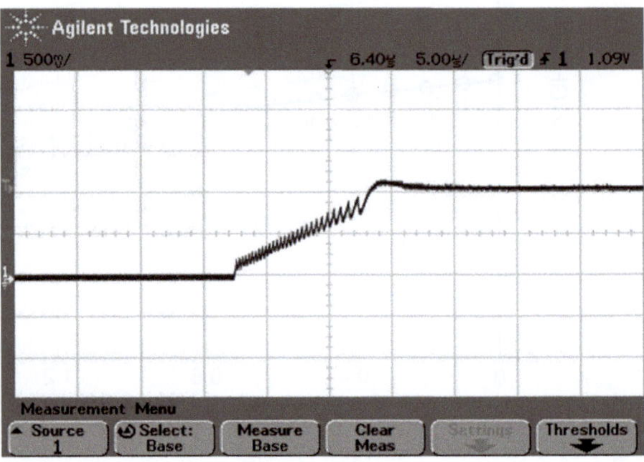

Table 3.1 Performance summary of the PLL

Technology	0.13 μm SiGe BiCMOS
Power Consumption	32 mA from 2.0~2.3 V supply
Output Frequency	4.8–6.8 GHz
Phase Noise	-86 dBc/Hz@10 kHz, -114 dBc/Hz@1MHz
Reference Spur	<-65 dBc
Loop bandwidth	100 kHz with 80 MHz F_{ref}
Lock Time	15 μs
Area (mm²)	1.35×0.65

References

1. B. Soltanian and P. Kinget, "A Tail Current-Shaping Technique to Reduce Phase Noise in LC VCOs," IEEE Custom Integrated Circuits Conference, pp. 579–582, 2005.
2. A. D. Berny, Ali M. Niknejad, and R. G. Meyer, "A 1.8-GHz LC VCO With 1.3-GHz Tuning range and Digital Amplitude Calibration," IEEE J. Solid-State Circuits, vol. 40, No. 4, pp. 909–917, Apr. 2005.
3. R. Nonis, E. Palumbo, P. Palestri, and L. Selmi, "A Design Methodology for MOS Current-Mode Logic Frequency Dividers," IEEE Transactions on Circuits and Systems-I: Regular papers, vol. 54, No. 2, pp. 245–254, Feb. 2007.
4. D. Lim, J. Kim, J. Plouchart, C. Cho, D. Kim, R. Trzcinski, and D. Boning, "Performance Variability of a 90 GHz Static CML Frequency Divider in 65 nm SOI CMOS," ISSCC Dig. Tech. Papers, pp. 542–543, Feb., 2007.
5. C. Vaucher, I. Ferencic, M. Locher, S. Sedvallson, U. Voegeli, and Z. Wang, "A Family of Low-Power Truly Modular Programmable Dividers in Standard 0.35-um CMOS Technology," IEEE J. Solid-State Circuits, vol. 35, No. 7, pp. 1039–1045, July 2000.
6. S. Pamarti, L. Jansson, and I. Galton, "A Wideband 2.4-GHz Delta-Sigma Fractional-N PLL With 1-Mb/s In-Loop Modulation," IEEE J. Solid-State Circuits, vol. 39, No. 1, pp. 49–62, Jan. 2004.
7. R. J. Wildlar, "New developments in IC voltage regulators," IEEE J. Solid-State Circuits, vol. 1, No. 1, pp. 2–7, Jan. 1971.
8. A. Fard and P. Andreani, "An Analysis of 1/f2 Phase Noise in Bipolar Colpitts Oscillators (With a Digression on Bipolar Differential-Pair LC Oscillators," IEEE J. Solid-State Circuits, vol. 42, No. 2, pp. 374–384, Feb. 2007.
9. B. Razavi, "Design of Analog CMOS Integrated Circuits," McGraw-Hill Science/Engineering/Math, Aug. 2000.

Chapter 4
Design and Analysis of QVCO with Various Coupling Techniques

4.1 Introduction

Quadrature signals are widely used in image-rejection transceivers [1] and half-rate clock and data recovery (CDR) systems [2]. Among the many quadrature-signal generation mechanisms [3–6], quadrature inductor–capacitor (LC) oscillator continues to be an attractive research topic since the first publication of an LC-QVCO in 1996 [7, 8], maily due to its excellent phase noise performance. Several quadrature techniques have been developed in search of improved phase noise performance, reduced phase accuracy, and elimination of bi-modal oscillation, such as top/bottom series coupling, transformer-coupling, capacitive-source coupling, and parallel coupling with phase shifters [7, 9–12]. On the other hand, coupling with active devices either introduces extra noise sources in the case of parallel coupling or lead to decreased voltage head room in the case of top/bottom series coupling. However, it is very challenging to design a QVCO that can completely remove the noise source in the quadrature coupling devices and avoid bi-modal oscillation simultaneously. A reliable QVCO structure should possess the following features:

a. A simple structure;
b. Introduces ignorable noise in the quadrature coupling path;
c. Provides deterministic quadrature outputs, i.e., $0°$ and $90°$;
d. No phase noise degradation due to quadrature coupling, namely, the QVCO phase noise performance should be close to or even better than its single-phase counterpart, i.e., $L_{QVCO}(\Delta f) = L_1(\Delta f) - 10 \log 2$;
e. Offers good phase accuracy over wide frequency tuning range.

F. Zhao, F. F. Dai, *Low-Noise Low-Power Design for Phase-Locked Loops,*
DOI 10.1007/978-3-319-12200-7_4

In order to design a robust QVCO that can achieve all those goals, innovative techniques have to be developed. In the following sections of this chapter, a quadrature LC VCO using capacitive-coupling technique combined with source degeneration capacitor is introduced to offer low phase noise performance, deterministic quadrature outputs, and excellent phase accuracy over wide frequency range. The utilization of source degeneration capacitor introduces intrinsic leading phase delay in the quadrature-coupling path, efficiently eliminating the phenomenon of bi-modal oscillation.

This chapter is organized as follows. Section 4.2 analyzes important aspects for a QVCO design. A capacitive-coupling QVCO with inherent leading phase shifter for the elimination of bi-modal oscillation is proposed in Sect. 4.3. Linear model and mode rejection ratio have been developed to evaluate the robustness of the proposed CC-QVCO structure. Discussed in Sect. 4.4 are the design details of the proposed CC-QVCO and corresponding simulation results to show its robustness. Besides, an SVCO counterpart and a class-C mode top-series (TS)-QVCO have been implemented for comparison. Experiment setup and measurement results are given in Sect. 4.5. Finally, conclusion is drawn to summarize this chapter.

4.2 Important Aspects of QVCO and Prior QVCO Structures

Figure 4.1 shows a classic QVCO [8] structure using parallel transistors for quadrature coupling. Each of the two VCO cores for quadrature generation consists of two cross-coupled transistors to provide negative g_m to the LC tank and another two transistors for quadrature coupling. The resonant frequency of a classic QVCO will deviate from the resonant frequency of its VCO core due to the quadrature coupling mechanism. The output phase relationships between the two ideal VCO outputs are ambiguous because the phase relationship can be either $+90°$ or $-90°$ [13]. Even though this ambiguity is relaxed for the reason that the practical asymmetry and parasitics usually lead to a unique solution to the oscillation conditions, phase directing circuits are often required to produce deterministic quadrature outputs.

A simplified linear model of the QVCO is shown in Fig. 4.2a, which includes the main transconductance G_{MA} to compensate the energy loss in the LC tank, G_{MC} for quadrature coupling, and LC tanks. A leading phase delay of θ is introduced in the quadrature coupling path G_{MC}. The voltage and current relationships for $+90°$ and $-90°$ conditions are shown in Fig. 4.2b. Current i_{AI} and i_{AQ} are generated by oscillation transconductance G_{MA} while i_{CI} and i_{CQ} are produced by the quadrature coupling transconductance G_{MC}. The signal amplitude of the VCO outputs can be easily obtained from the phasor diagram as

$$\begin{cases} \text{Model} : A_{m1} = \alpha R_P \left(i_{AI} + i_{CI}\sin\theta \right) = \alpha R_P i_{AI} \left(1 + m\sin\theta \right) \\ \text{Mode2} : A_{m2} = \alpha R_P \left(i_{AI} - i_{CI}\sin\theta \right) = \alpha R_P i_{AI} \left(1 - m\sin\theta \right) \end{cases} \tag{4.1}$$

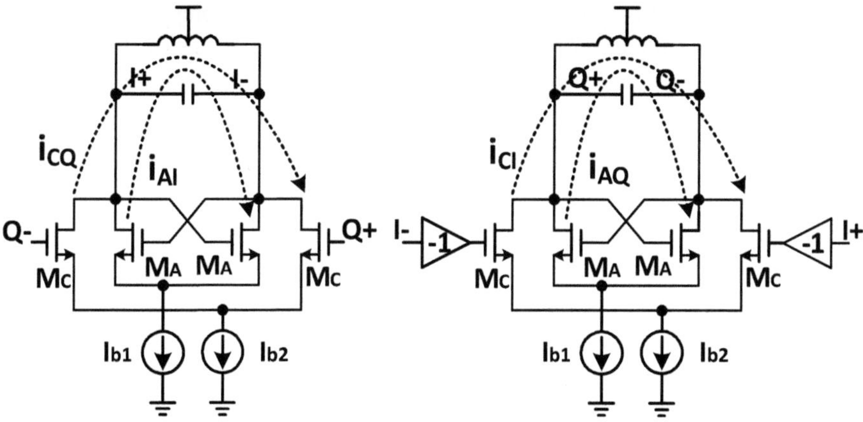

Fig. 4.1 Classic QVCO structure utilizing parallel coupling

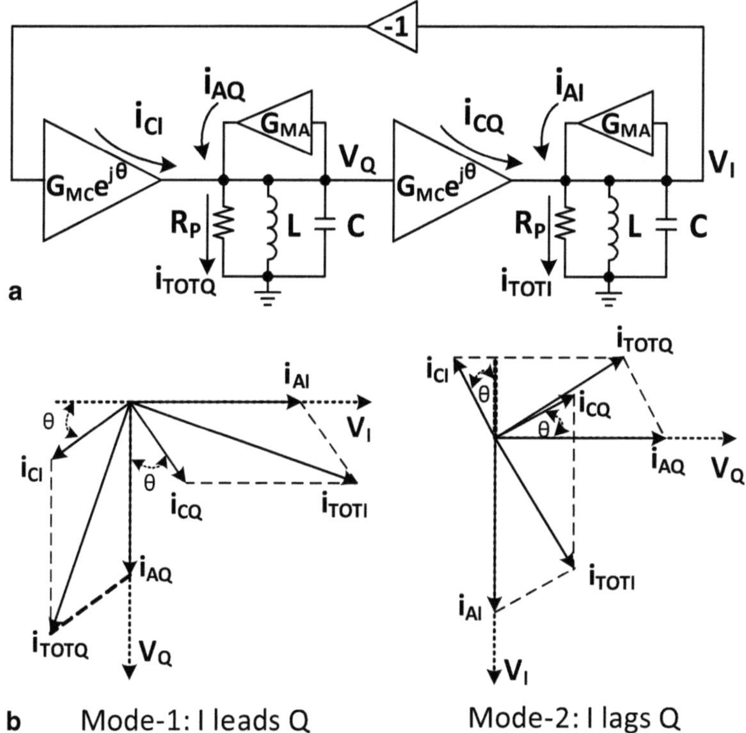

Fig. 4.2 **a** Linear model of conventional QVCO; and **b** phasor diagram illustration of voltage and current for two phase relationships

where $\alpha=2/\pi$ for all NMOS or PMOS VCO and $\alpha=4/\pi$ for complementary VCO in current limited region, R_p is the equivalent parallel resistance of the LC tank, and m is the quadrature-coupling factor defined as the current ratio of i_{CI} to i_{AI}. For the QVCO structure shown in Fig. 4.1, $\alpha=2/\pi$, $i_{AI}=i_{AQ}=I_{b1}$, and $i_{CI}=i_{CQ}=I_{b2}$. When the phase delay θ is 0, the QVCO can be in either mode 1 or 2. Perturbation analysis of the QVCO show that mode 2 is conditionally stable when $\sin(\theta)<m$ while mode 1 is unconditionally stable [11]. Note that the phase delay in reference [11] is a lagging delay while the delay in this chapter is positive, but the above analysis is still valid. In order to avoid phase ambiguity, it is necessary to force the QVCO operating in mode 1, and thus the phase delay should meet the condition of $\theta>\arcsin(m)$.

On the other hand, the phase noise performance of the QVCO with parallel coupling is usually deteriorated because of the limited delay in the quadrature-coupling path. Assuming tail current is noiseless, the phase noise in thermal noise region for a mode-1 QVCO can be expressed as [11]

$$L(\Delta\omega) = 10\log\left[\left(kTR_P/2V_{QVCO}^2\right)F_{QVCO}\left(\omega_0/Q\Delta\omega\right)^2\right]$$

$$F_{QVCO} = 1+\underbrace{\left(m\cos\theta/1+m\sin\theta\right)^2}_{\text{tank noise}}+\gamma\left[\underbrace{\left(1/1+m\sin\theta\right)}_{\text{MA noise}}\left(1+\underbrace{m\left(m+\sin\theta/1+m\sin\theta\right)^2}_{\text{MC noise}}\right)\right]$$

$$(4.2)$$

where k is Boltzmann constant, T is absolute temperature, V_{QVCO} is the signal amplitude of the output signal, Q is the quality factor of the LC tank, ω_0 is the resonant frequency, $\Delta\omega$ is the frequency offset, and γ is the body effect parameter. The phase noise of a SVCO in thermal noise region can be expressed as

$$L(\Delta\omega) = 10\log\left[(kTR_P/V_{SVCO}^2)F_{SVCO}\left(\omega_0/Q\Delta\omega\right)^2\right]$$
$$F_{SVCO} = 1+\gamma$$
$$(4.3)$$

where V_{SVCO} is the signal amplitude of the output signal and $V_{QVCO} = (1+m\sin\theta)V_{SVCO}$.

The phase noise of QVCO compared with its SVCO counterpart is shown in Fig. 4.3. When the phase delay θ is small, the phase noise of QVCO increases as the quadrature-coupling factor m goes up. It is better to reduce the coupling factor m under this condition. Typically, a conventional QVCO with parallel coupling shows a phase noise performance of 4–6 dB worse than its SVCO counterpart. The situation changes to the opposite when phase delay is higher than $20°$ and the phase noise improves as m increases. The improvement becomes more obvious when the phase delay is above $40°$. Therefore, stronger coupling factor is preferred for larger phase delay. However, there is a boundary defined by $\theta=\arcsin(m)$ which separates the stable region and ambiguous region. In order to avoid phase ambiguity, it is desirable to choose m and phase delay below the dashed boundary line shown in Fig. 4.3. The triangular gray box in Fig. 4.3 shows the optimum region where it has better noise performance than its single-phase counterpart and deterministic outputs.

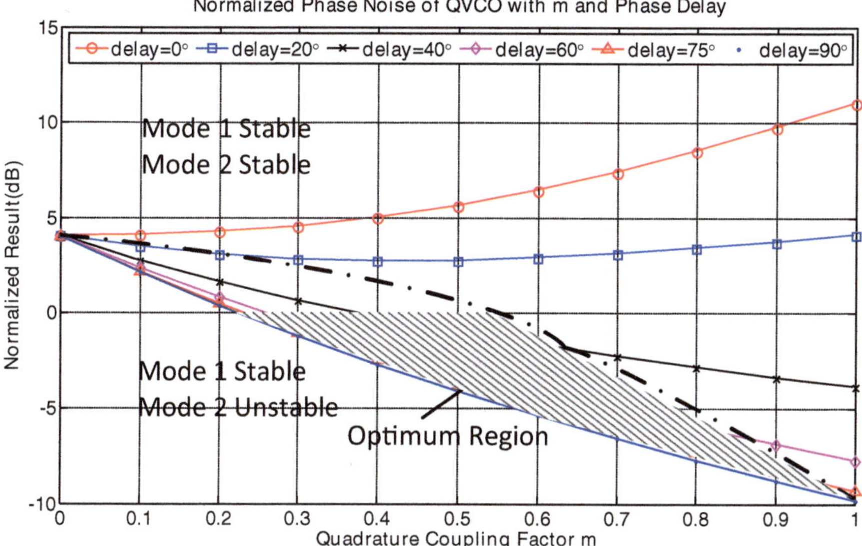

Fig. 4.3 QVCO phase noise normalized to SVCO noise for different *m* and phase delay (phase noise of SVCO is at 0 dB), and γ is assumed to be 2 for short-channel MOS transistors

The phase noise performance of a conventional QVCO structure with $\theta = 0$ is worse than its single-phase counterpart due to not only the quadrature coupling but also the additional noise source in the quadrature-coupling path. Without the noise source in the quadrature-coupling device, the phase noise in thermal noise region can be reduced to

$$L(\Delta\omega) = 10\log\left[(kTR_{\mathrm{P}} / 2V_{\mathrm{QVCO}}^2)\left(1 + \underbrace{(m\cos\theta / 1 + m\sin\theta)^2}_{\text{tank noise}} + \underbrace{(\gamma / 1 + m\sin\theta)}_{\text{MA noise}} \right)(\omega_0 / Q\Delta\omega)^2 \right]$$

$$(4.4)$$

The QVCO phase noise performance without quadrature-coupling devices MC is lower than that with MC. The normalized phase noise results are plotted in Fig. 4.4. The noise improvement converges to 4 dB as coupling factor increases to one. From previous analysis, it is well known that the phase noise performance becomes worse when the coupling factor of a conventional QVCO without phase delay in quadrature-coupling path increases. Therefore, it is desirable to develop quadrature-coupling techniques to eliminate the noise degradation in the quadrature-coupling path.

4.3   CC-QVCO with Leading Phase Delay

This section introduces a capacitive-coupled QVCO (CC-QVCO) with embedded leading phase shifter as shown in Fig. 4.5 and it has the following features: low phase noise performance, no bi-modal oscillation, and good phase accuracy across

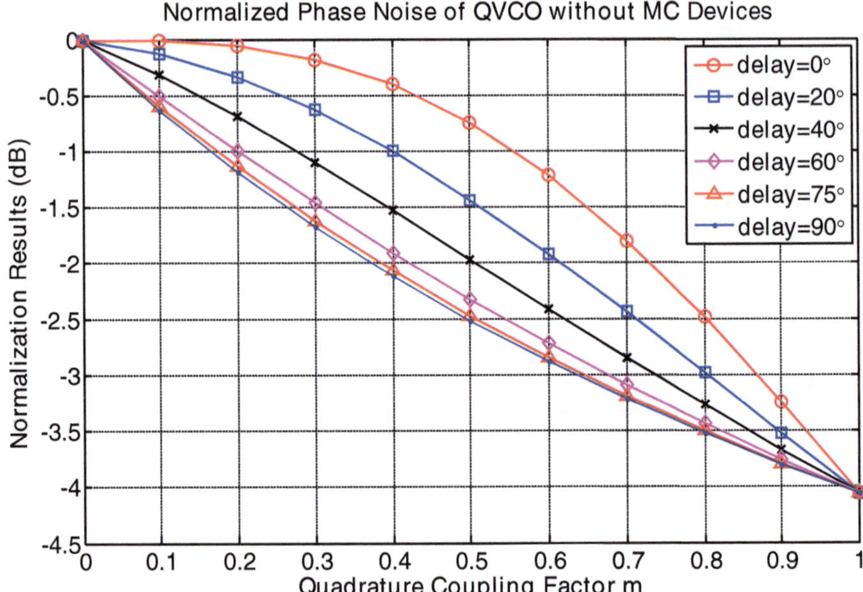

Fig. 4.4 QVCO phase noise without MC devices normalized to QVCO noise with MC devices for different m and phase delay, and γ is assumed to be 2 for short-channel MOS transistors

wide frequency tuning range [14]. In order to improve the phase noise performance, capacitive coupling mechanism is adopted to form quadrature coupling. The cross-coupling capacitors C_{CC} and the transistors provide negative-g_m to compensate the energy loss in the LC tank while the quadrature-coupling capacitors C_{QC} couple the

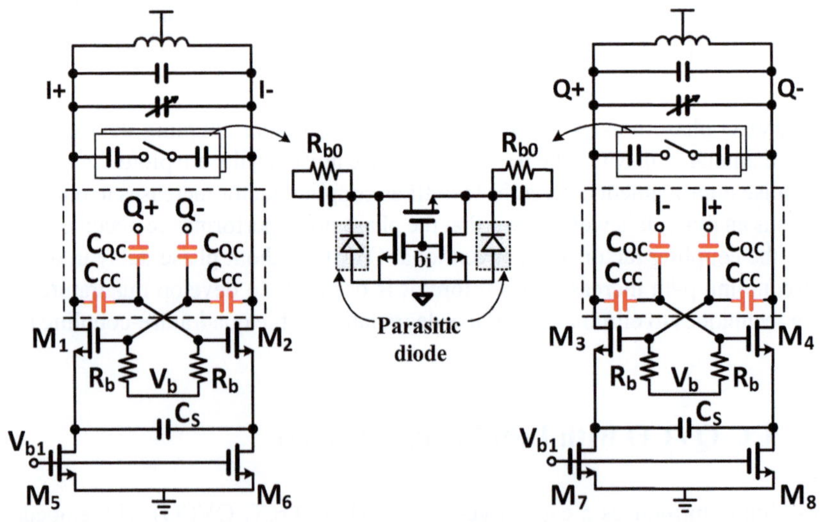

Fig. 4.5 Proposed CC-QVCO with embedded leading phase shifter for quadrature signal generation

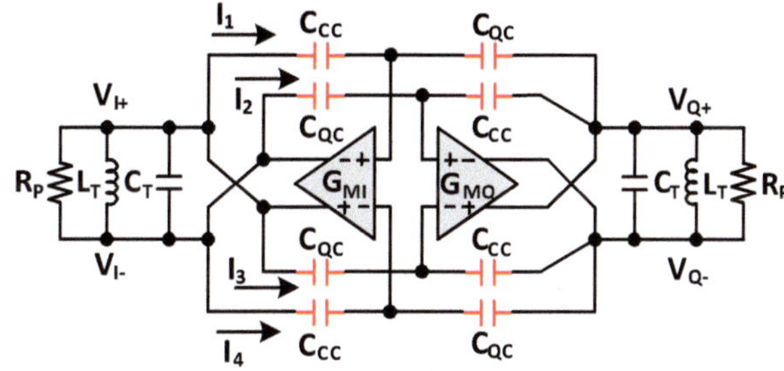

Fig. 4.6 Linear model of QVCO including the quadrature-coupling capacitors and cross-coupling capacitors

two VCO cores to allow quadrature signal generation. With the capacitive quadrature-coupling technique, the impulse sensitivity function which has been used to evaluate the phase noise performance can be improved for a Colpitts QVCO [12]. By completely removing the noisy quadrature-coupling transistors in conventional parallel-coupling QVCOs, the phase noise performance of the proposed CC-QVCO should be improved. Moreover, to avoid the problem of phase ambiguity, source-degenerated VCO core is utilized to provide leading phase delay in the quadrature-coupling path. Metal-insulator-metal (MIM) capacitor array with 2-bit control is used to provide the coarse frequency tuning. The frequency tuning range has been extended by employing a large resistor to reversely bias the parasitic diode in the NMOS switches.

4.3.1   *Linear Model of the CC-QVCO and Start-up Conditions*

The operation of the CC-QVCO core is different from conventional parallel coupling QVCO since both the cross-coupling signal and the quadrature-coupling signal will be coupled to the gates of the negative-g_m transistors. The resonant frequency of the CC-QVCO is also different from a classic QVCO because the LC tanks are loaded with additional capacitors used for quadrature coupling and cross coupling. Figure 4.6 shows the linear model of the CC-QVCO including the loading effect of capacitors C_{CC} and C_{QC}. The loading effect of the coupling capacitors can be represented as an equivalent capacitor loading to each of the QVCO tank. It can be found by calculating the current flowing through those capacitors. By defining $C_X = C_{QC} \| C_{CC}$ the currents shown in Fig. 4.6 are expressed as

$$I_1 = (V_{I+} - V_{Q+})sC_X \tag{4.5}$$

$$I_2 = (V_{I-} - V_{Q+})sC_X \tag{4.6}$$

$$I_3 = (V_{I+} - V_{Q-})sC_X \tag{4.7}$$

$$I_4 = (V_{I-} - V_{Q-})sC_X \tag{4.8}$$

Assuming that the output signals are differential, i.e. $V_{I+} = -V_{I-}$, $V_{Q+} = -V_{Q-}$, the equivalent capacitor loading to the tank is

$$C_C = \frac{I_1 + I_3}{s(V_{I+} - V_{I-})} = \frac{(V_{I+} - V_{Q+})sC_X + (V_{I+} - V_{Q-})sC_X}{s(V_{I+} - V_{I-})} = C_X \tag{4.9}$$

Thus, the total tank capacitance becomes $C = C_T + C_X$. With superposition theorem, the current flowing out of the left transconductance G_M can be expressed as

$$\overline{I_{MI}} = \frac{C_{CC}}{C_{CC} + C_{QC}} G_M \left(V_{I+} - V_{I-}\right) + \frac{C_{QC}}{C_{CC} + C_{QC}} G_M \left(V_{Q+} - V_{Q-}\right) \tag{4.10}$$

Therefore, the effect of transconductance can be expressed with two terms: the first term on the right hand of the above equation represents the effect of negative-g_m used to start the oscillator, and the second term represents the effect of quadrature coupling for quadrature signal generation. Then, an equivalent linear model for the proposed CC-QVCO can be developed as shown in Fig. 4.7. The leading phase delay introduced by the source degeneration capacitor is modeled as θ in the range of $[0, 90°]$.

Assuming the bias resistor R_b is sufficiently large and its effect on the phase shifter can be neglected, the effective cross-coupling transconductance and quadrature-coupling transconductance can be derived as

$$\overline{G_{MA}} = \frac{C_{CC}}{C_{CC} + C_{QC}} \frac{g_{m1,2}sC_S}{0.5g_{m1,2} + sC_S} = G_{MA}e^{j\theta} \tag{4.11}$$

$$\overline{G_{MC}} = \frac{C_{QC}}{C_{CC} + C_{QC}} \frac{g_{m1,2}sC_S}{0.5g_{m1,2} + sC_S} = G_{MC}e^{j\theta} \tag{4.12}$$

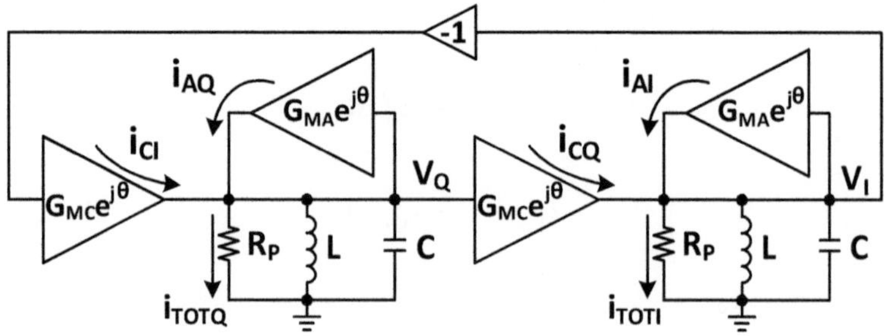

Fig. 4.7 Simplified linear model of the proposed QVCO with phase shifter

where $\theta = 90° - \arctan\left(2\omega_0 C_S / g_{m1,2}\right)$. Quadrature-coupling strength factor m is redefined as

$$m = \frac{G_{MC}}{G_{MA}} = \frac{C_{QC}}{C_{CC}} \tag{4.13}$$

Furthermore, the loop gain based on the linear model is given by

$$G(\omega) = -\left(\frac{G_{MC}e^{j\theta}}{(1/R_P)+(1/j\omega L) + j\omega C - G_{MA}e^{j\theta}}\right)^2 \tag{4.14}$$

where the real part and imaginary part of the term in the parentheses can be written as

$$\text{Real} = \frac{G_{MC}\cos\theta((1/R_P)-G_{MA}\cos\theta)+G_{MC}\sin\theta(\omega C-(1/\omega L)-G_{MA}\sin\theta)}{((1/R_P)-G_{MA}\cos\theta)^2+(\omega C-(1/\omega L)-G_{MA}\sin\theta)^2} \tag{4.15}$$

$$\text{Imag} = \frac{G_{MC}\sin\theta((1/R_P)-G_{MA}\cos\theta)-G_{MC}\cos\theta(\omega C-(1/\omega L)-G_{MA}\sin\theta)}{((1/R_P)-G_{MA}\cos\theta)^2+(\omega C-(1/\omega L)-G_{MA}\sin\theta)^2} \tag{4.16}$$

For sustainable oscillation, Barkhausen's phase criteria should be met. Therefore, Eq. (4.15) should equal 0 and it leads to

$$\omega C - (1/\omega L) - G_{MA}\sin\theta = -\frac{\cos\theta}{\sin\theta}((1/R_P)-G_{MA}\cos\theta) \tag{4.17}$$

The derivation of the above equation is based on the assumption that $\theta \neq 0$. Inserting the above equation into Eq. (4.16), the absolute value of the imaginary part can be simplified as

$$\left|\text{Imag}\right| = \left|\frac{G_{MC}\sin\theta}{\dfrac{1}{R_P}-G_{MA}\cos\theta}\right| \tag{4.18}$$

Under the oscillation condition, the absolute value of the imaginary part equals to 1. The condition for Barkhausen's amplitude criteria is obtained as

$$\begin{cases} (G_{MA}\cos\theta+G_{MC}\sin\theta)R_P = 1, & 1/R_P - G_{MA}\cos\theta > 0 \\ (G_{MA}\cos\theta-G_{MC}\sin\theta)R_P = 1, & 1/R_P - G_{MA}\cos\theta < 0 \end{cases} \tag{4.19}$$

With Eq. (4.17), the oscillation frequency of the CC-QVCO can be solved as

$$\omega_{osc} = -\frac{\cos\theta / R_P - G_{MA}}{2C\sin\theta} + \sqrt{((\cos\theta / R_P - G_{MA})/2C\sin\theta)^2 + \omega_0^2} \quad (4.20)$$

where $\omega_0 = 1/\sqrt{LC}$. Under the condition of $1/R_P - G_{MA}\cos\theta > 0$, the first term of right-hand side can be simplified with the approximation equation of $Q = R_P\omega_0 C$, namely,

$$\Delta\omega_1 = -\frac{\omega_0}{2Q}\frac{-G_{MA}\sin\theta + G_{MC}\cos\theta}{G_{MA}\cos\theta + G_{MC}\sin\theta} = \frac{\omega_0}{2Q}\frac{\sin\theta - m\cos\theta}{\cos\theta + m\sin\theta} \quad (4.21)$$

Note that when the quality factor of the tank is much larger than 1, the first term of the inner square root of Eq. (4.20) is much smaller than the $\Delta\omega_1$ term and can be neglected. Then, the resonant frequency is approximated as

$$\omega_{osc1} = \omega_0 + \frac{\omega_0}{2Q}\frac{\sin\theta - m\cos\theta}{\cos\theta + m\sin\theta} \quad (4.22)$$

Similarly, the frequency deviation and oscillation frequency for the other condition of $1/R_P - G_{MA}\cos\theta < 0$ can be solved as

$$\Delta\omega_2 = -\frac{\omega_0}{2Q}\frac{-G_{MA}\sin\theta - G_{MC}\cos\theta}{G_{MA}\cos\theta - G_{MC}\sin\theta} = \frac{\omega_0}{2Q}\frac{\sin\theta + m\cos\theta}{\cos\theta - m\sin\theta} \quad (4.23)$$

$$\omega_{osc2} = \omega_0 + \frac{\omega_0}{2Q}\frac{\sin\theta + m\cos\theta}{\cos\theta - m\sin\theta} \quad (4.24)$$

Even though the special condition with $\theta \neq 0$ assumed in the above derivation, the analytical results also apply to the special condition of $\theta = 0$ and the resonant frequency under such condition is $\omega_{osc} = \omega_0 \pm m\omega_0 /(2Q)$.

4.3.2   Mode Rejection for Stable Operation

The proposed CC-QVCO with inherent leading phase delay also has two stable modes associated with different quadrature phase sequence. It is hazardous to use a QVCO with ambiguous quadrature outputs to drive an image rejection receiver, because it is unclear whether the upper or lower sideband will be selected. Although corrections can be done in baseband with extra circuits, it is still desirable to generate quadrature LOs without any phase ambiguity. The voltage and current relationship in the CC-QVCO is different from that of a conventional QVCO because both the quadrature-coupling transconductance and the negative-g_m stage have a leading phase delay of θ. It is thus helpful desirable to understand the effect of the phase delay on the stability of the CC-QVCO.

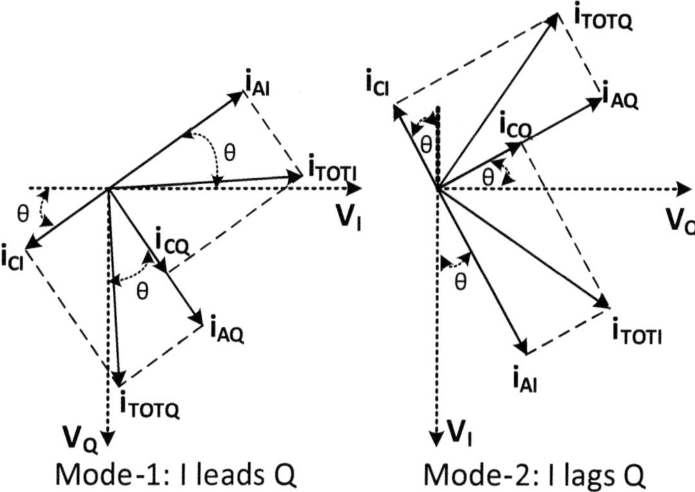

Fig. 4.8 Phasor diagram of the voltage and current relationships in the proposed oscillator

Assuming the quadrature output voltages as $V_I(t) = A\cos(\omega_0 t)$ and $V_Q(t) = A\cos(\omega_0 t - \varphi)$ with initial phase of 0, the voltage signals can be expressed in exponential form as $V_I = Ae^{j0}$ and $V_Q = Ae^{-j\varphi}$. Figure 4.8 illustrates the voltages and currents in the proposed QVCO. The signal amplitudes corresponding to the two stable modes can be expressed as

$$\begin{cases} \text{Mode1}: A_{m1} = \alpha R_P \left(i_{AQ}\cos\theta + i_{CI}\sin\theta \right) = \alpha R_P i_{AQ} \left(\cos\theta + m\sin\theta \right) \\ \text{Mode2}: A_{m2} = \alpha R_P \left(i_{AQ}\cos\theta - i_{CI}\sin\theta \right) = \alpha R_P i_{AQ} \left(\cos\theta - m\sin\theta \right) \end{cases} \quad (4.25)$$

where m is defined by Eq. (4.13) and α is a coefficient modeling the effect of VCO topology and source degeneration.

In practical applications, it is difficult to derive an accurate phase delay required to completely eliminate the problem of bi-modal oscillation due to the asymmetry in the two VCO cores since it is very challenging to develop an accurate QVCO model to include the nonlinear effect resulted from large-signal operation. However, it is possible to find a robust solution which shows high rejection to the unwanted mode. A mode rejection ratio (MRR) representing a QVCO's capability to reject the unwanted operation mode can be simulated.

First, we can enhance the start-up gain for the unwanted mode by injecting energy into the LC tank. The unwanted mode, assuming to be $-90°$ mode, will start to oscillate with the help of the artificially introduced energy. After the signal amplitude for $-90°$ mode reaches to some value A_U, the energy injected can be removed immediately or damped with high damping ratio. Then, a robust QVCO should be able to attenuate the $-90°$ mode oscillation and gradually converge to $+90°$ mode with signal amplitude of A_W. It requires the QVCO to be able to provide higher

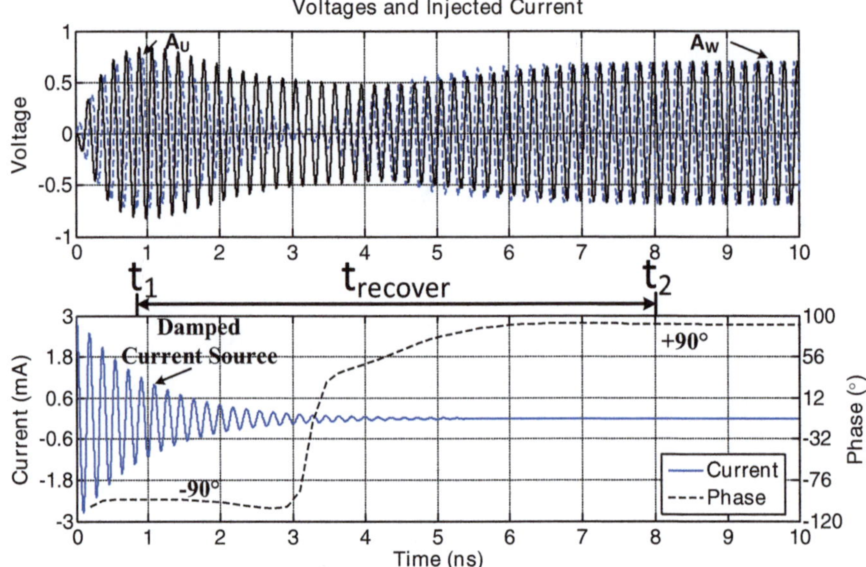

Fig. 4.9 Transient waveforms for MRR calculation

loop gain for $+90°$ mode than $-90°$ mode. The stronger the unwanted oscillation that an oscillator can reject, the more disturbances a QVCO can tolerate for stable operation. According to the above analysis, the mode rejection ratio is defined by the following equation:

$$MRR = 20log\frac{A_U}{A_W}.$$ (4.26)

Figure 4.9 illustrates all the parameters used for the calculation of mode rejection ratio. By injecting two damped quadrature currents with $-90°$-phase relationship into the two LC tanks, respectively, the QVCO will start oscillating in $-90°$ mode from $t=0$. The $-90°$ mode signal amplitude reaches its maximum value at t_1. In the meantime, the quadrature currents have been damped to a relatively small amplitude. A robust QVCO will be able to move away from $-90°$ from t_1 and then stabilize its $+90°$ mode oscillation at t_2. During the time period from t_1 to t_2, the QVCO gradually converges to the wanted mode. The waveforms shown in Fig. 4.9 have the following parameter: $A_U = 0.73$ V, $A_W = 0.7$ V, $T_{VCO}=182$ ps, and $t_{recovery}=7.2$ ns and

$$MRR_{example} = 20log\frac{0.73V}{0.7V} = 0.36dB$$ (4.27)

It means that the QVCO is able to tolerate an unwanted oscillation mode that is at least 0.36 dB higher than the wanted mode. The higher the MRR is, the better the rejection to the unwanted disturbance will be. In order to find out the maximum MRR for a specific QVCO structure, the injecting energy needs to be gradually tuned to reach the boundary condition.

4.4   Design and Analysis of a 5.6 GHz CC-QVCO

4.4.1   Design Procedure of a CC-QVCO

The primary goal of a QVCO design is to provide deterministic quadrature outputs, low phase noise performance, small phase accuracy, and tolerance to PVT variations. In order to demonstrate the robustness of the capacitive-coupling technique, an example of CC-QVCO with leading phase shifter is implemented in a 130 nm CMOS process. Symmetrical spiral inductor with a size of 220×220 μm^2, realized on a 4-μm-thick top metal, is chosen for the resonant tank. The quality factor of the inductor plays a key role in achieving low phase noise performance and thus should be optimized by adjusting the space, metal width, and outer dimension. The quality factor of the inductor has been simulated for different combinations of device sizes at the targeted frequency. The differential quality factor and inductance extraction used for inductor optimization is obtained from the S-parameters of two-port simulation results defined by the following equations [15]:

$$S_{\text{diff}} = \frac{S_{11} + S_{22} - S_{12} - S_{21}}{2} \tag{4.28}$$

$$Z_{\text{diff}} = 2R_0 \frac{1 + S_{\text{diff}}}{1 - S_{\text{diff}}} \tag{4.29}$$

$$Q_{\text{diff}} = \frac{\text{Imag}(Z_{\text{diff}})}{\text{Real}(Z_{\text{diff}})}, L_{\text{diff}} = \frac{\text{Imag}(Z_{\text{diff}})}{\omega} \tag{4.30}$$

where R_0 is the default single-ended port impedance. After optimization, the quality factor of a 2-turn, 1. 12nH inductor is 16 with metal space of 5 μm and metal width of 15 μm. Moreover, the quality factor of the LC tank including the varactor and capacitor array is also optimized to achieve the best noise performance. Two-bit MIM capacitor array is used to provide a coarse tuning range of 1 GHz. In order to minimize the degradation of quality factor, the switch with minimum length and width of 60 μm is used for coarse frequency tuning. According to reference [9], a mutual inductance M between the tanks of the two VCO cores exists. The strength of the coupling factor M is dependent on the distance of the two inductors. In order to reduce the mutual inductance that will deteriorate the phase error of the quadrature outputs, the distance between the inductors is chosen to be 500 μm. Other components such as transistors and capacitor arrays are placed between the inductors. The startup condition can be derived from Eqs. (4.11)–(4.13) and (4.19). Considering both the phase error and phase noise performance, $m=0.6$ and phase delay $\theta=30°$ are chosen to implement the proposed CC-QVCO.

4.4.2   Phase Noise and Phase Error with Current Bias

It is well known that a cross-coupled single-phase LC VCO with current tail allows two regime of operation [16]: (a) current-limited region where the tank amplitude is solely determined by the tail-current source and the tank equivalent resistance; (b) voltage-limited region where the tank amplitude is clipped by the supply voltage V_{DD} and the tail current is not constant since the tail transistor is in the triode region. Usually a VCO is optimized to operate at the edge between the current-limited region and voltage-limited region within a given power dissipation. However, the situation for a QVCO design is a little different from conventional single-phase VCO design because the phase error performance also depends on the operating region. Consequently, the optimum operation region for QVCO design should be determined in order to achieve minimum phase noise and phase error performance.

First let us find the signal amplitude and phase noise with the bias current in a single-phase VCO. The oscillator used to evaluate the amplitude and noise is the VCO core used to build the proposed QVCO as shown in Fig. 4.10. The signal amplitude in the LC tank is determined by the following equation [16]:

$$V_{\text{SVCO}} = \frac{2}{\pi} I_B R_P \tag{4.31}$$

where I_B is total current flowing through the two tail transistors and R_P is the equivalent tank resistance. According to Leeson's equation, the VCO phase noise is inversely proportional to the signal amplitude in the LC tank [17]. It is desirable to increase the signal as much as possible to achieve low phase noise performance.

The simulated signal amplitude and phase noise performance of the single-phase VCO core for different bias conditions are shown in Fig. 4.11. The output signal amplitude increases monotonically with increase of the tail current bellow 2 mA, which can be defined as current-limited region and the slop is around 360 mV/mA. Further increasing the bias current can increase the signal amplitude, yet the phase noise performance first becomes worse and then drops at the expense of an increased power consumption. Thus, the optimum operation condition from noise perspective is to bias the VCO around the noise minima where both the phase noise and FoM will be optimum, i.e., 1.6 mA current bias as shown in Fig. 4.11.

Similar to a single-phase VCO, the amplitude of the output signal of a QVCO also increases when the bias current grows up. Although the signal amplitude in-

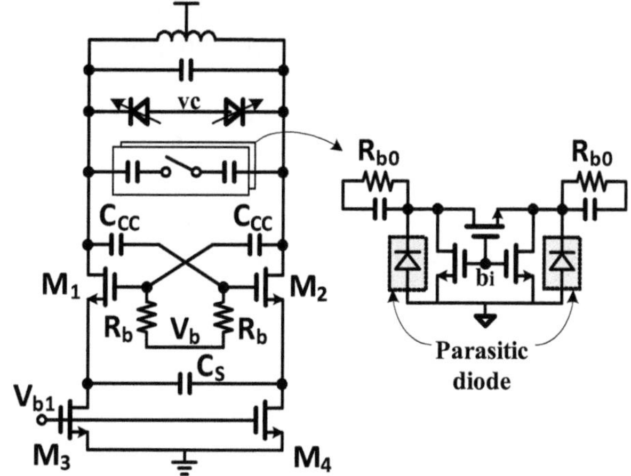

Fig. 4.10 Single-phase VCO (SVCO) for comparison

creases with bias current in the current-limited region (at a slop of 320 mV/mA for half QVCO current), the minimum phase noise achievable is -122.8 dBc/Hz @ 1 MHz offset. The phase noise will become worse if the bias current increases beyond 2.1 mA for 1.2 V supply voltage. Larger coupling factor leads to slightly better phase noise performance. On the other hand, the quadrature phase accuracy slowly degrades in the current-limited region where the phase noise improves with the increment of the bias current. After the current grows beyond the point corresponding to the noise minima, the phase error deviates from 90° rapidly, especially

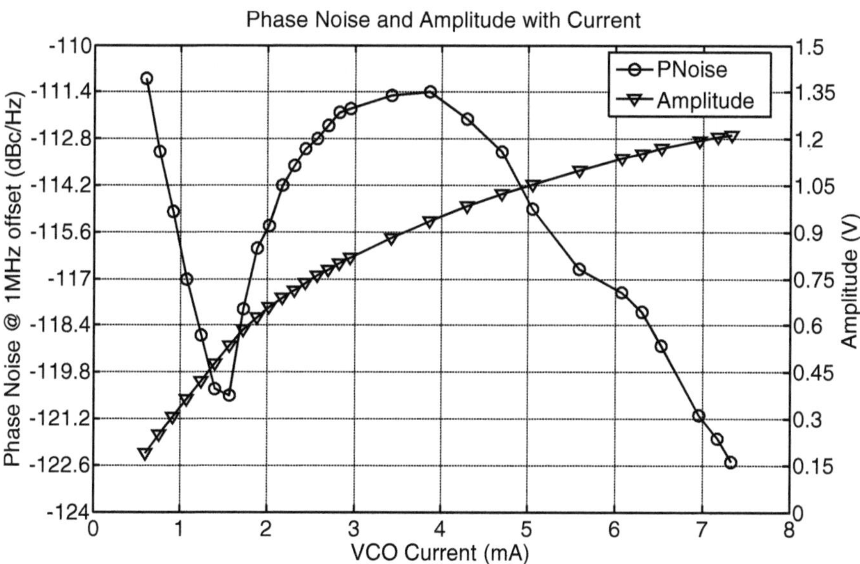

Fig. 4.11 Simulated signal amplitudes and phase noise for SVCO

for $m=0.6$. The quadrature phase accuracy for $m=1$ is much more stable than that for $m=0.6$, but it consumes much more power. From the noise simulation results for the QVCO and SVCO, it can be observed that the minimum noise for the proposed QVCO is about 2.3 dB better than its SVCO core while the theory prediction is 3 dB.

4.4.3   Class-C Mode TS-QVCO for Comparison

A TS-QVCO structure [18], featuring the low noise, small area cost, good suppression of 1/f noise, and compact implementation, is implemented for comparison. It uses top-series transistors whose noise can be degenerated because of the cascode configuration to form the quadrature coupling. Therefore, the phase noise performance of a TS-QVCO can be as good as its single-phase counterpart. In order to further improve the phase noise performance, a TS-QVCO with tail-current shaping technique is implemented for comparison. Moreover, AC coupling capacitors are utilized to implement the cross coupling for each VCO core. The combination of the AC coupling capacitor and degenerating capacitor C_S allows the VCO to operate in class-C mode, which has been demonstrated with better phase noise performance than a conventional cross-coupled VCO [19]. The sizes of the inductors and capacitor array for the LC tank in the proposed class-C mode TS-QVCO are the same as the one used for the proposed CC-QVCO. In this implementation, a quadrature-coupling factor $m_{TS-Q}=2$ is used to implement the proposed TS-QVCO.

Figure 4.12 shows the simulation results of the phase noise and signal amplitude of the class-C mode TS-QVCO. In the linear region, the output signal amplitude

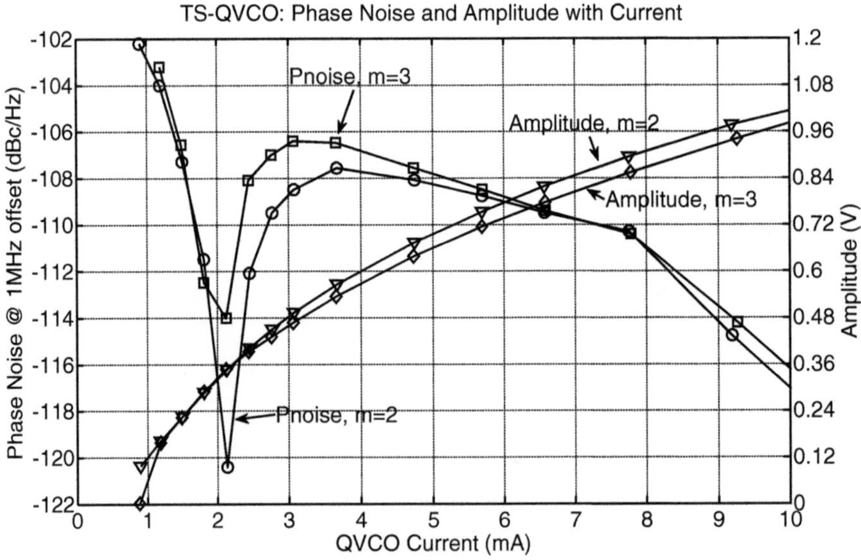

Fig. 4.12 Simulated phase noise and output amplitude of the proposed TS-QVCO

increases with a slope around 400 mV/mA assuming half QVCO current. The phase noise at 1-MHz offset first decreases to -120.3 dBc/Hz and then quickly rises to a high level. After its peak at 3.7 mA, the phase noise performance slowly improves with the sacrifice of the increased power consumption.

4.4.4   Performance Tolerance to Voltage and Temperature Variations

The practical environment for an integrated chip is usually complicated since the ambient temperature and supply voltage may change for a portable device. The current bias also also varies under different temperature or supply voltage conditions. Moreover, the performance of a circuit is also affected by the fabrication process. Therefore, a robust QVCO should be able to provide relatively constant performance over those variations. In order to prove the robustness of the proposed CC-QVCO, the variations of the phase noise and phase error with voltage and temperature variations are discussed in this section. Several simulations were run with spectreRF to compare the phase noise and phase error for the proposed CC-QVCO and class-C mode TS-QVCO. The phase noise for the CC-QVCO and its SVCO is also compared to show the robustness of the capacitive coupling technique.

4.4.5   Impact of Bias Current and Supply

Typically the current bias of a QVCO is optimized for only a few target applications; for example, minimum phase noise can be controlled by a few control bits for different bands. However, the current changes with component accuracy, ambient temperature, and fabrication process as it requires complex automatic control module to find the optimal setup for different frequency bands. Therefore, the robustness of QVCO structure highly relies on its tolerance to the current and supply change.

Figures 4.13 and 4.14 show the simulated performance variation with bias and supply voltage for a CC-QVCO and a TS-QVCO, respectively. The noise dependence on supply voltage for the CC-QVCO is higher than that of the TS-QVCO. The minimum achievable phase noise for the former is about 2–3 dB better than the latter but at the cost of 70 % more power consumption. The reason behind this extra power required to reach the noise minima is intuitive because the voltage headroom for TS-QVCO has been degraded by the top-series transistor. Therefore, the best FoM considering only power and phase noise for the two structures are close to each other. But the phase noise variation over $\pm 50\%$ current range is less than 5 dB for CC-QVCO while that for TS-QVCO is more than 10 dB.

The phase error for CC-QVCO is also relatively flattened around the minimum noise point but it starts to deteriorate for TS-QVCO. With the same mismatch in the LC tank and 1.2-V power supply, the phase error around the noise minima is $1.5°$

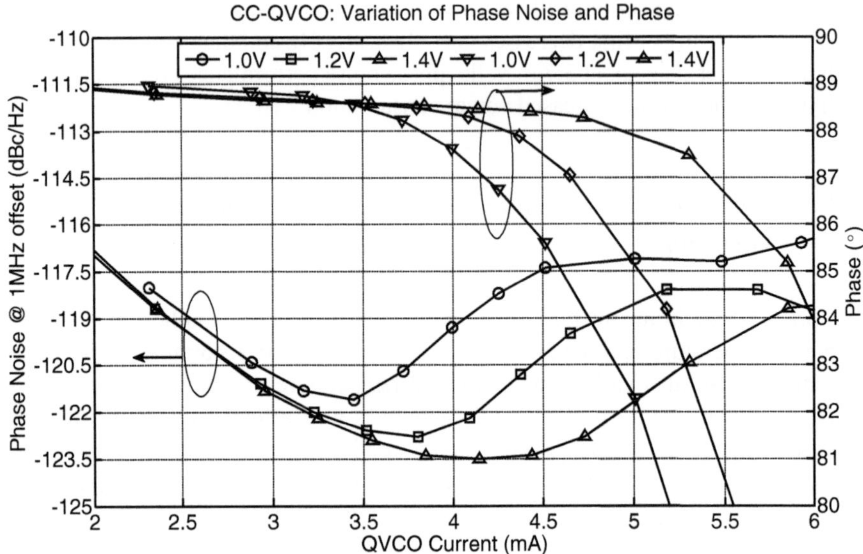

Fig. 4.13 CC-QVCO: simulated phase noise and phase error with bias and supply voltage (1 % capacitor mismatch was manually added to the LC tank)

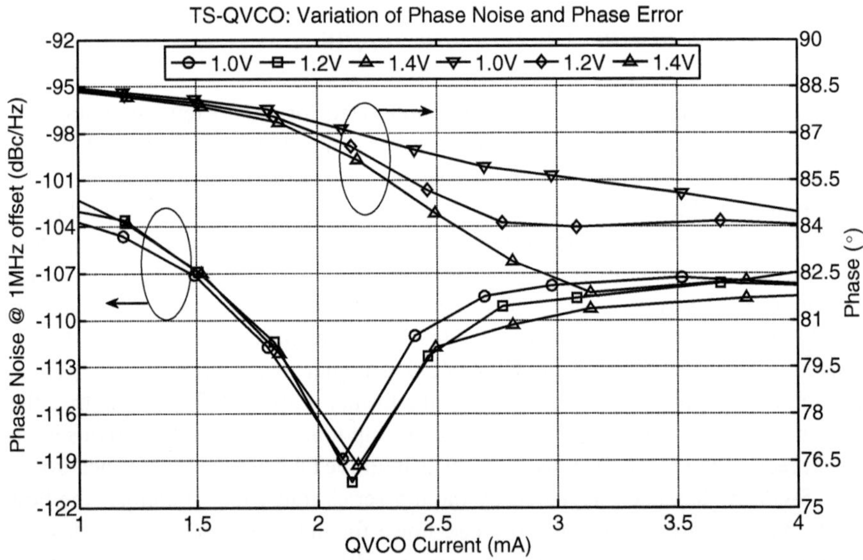

Fig. 4.14 TS-QVCO: simulated phase error and phase noise with bias and supply voltage (1 % capacitor mismatch was manually added to the LC tank)

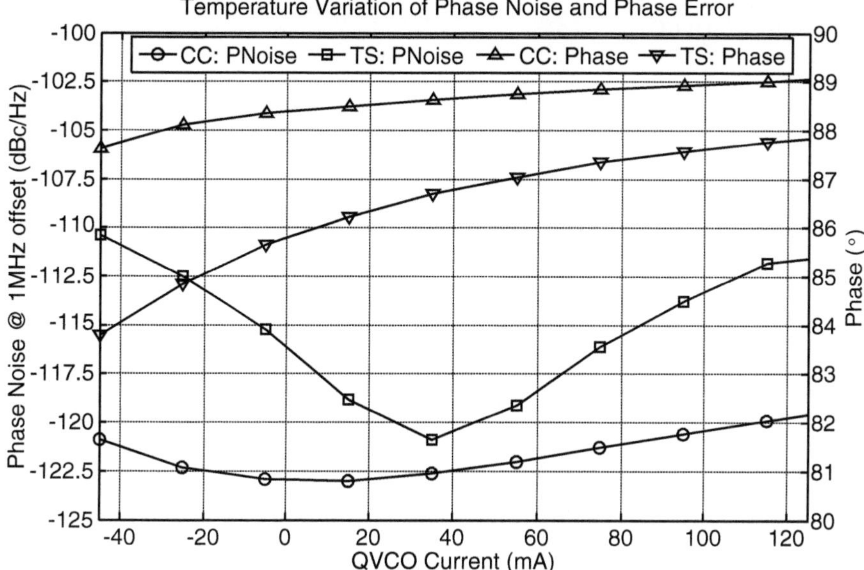

Fig. 4.15 QVCO phase error and phase noise with temperature (1 % capacitor mismatch was manually added to the LC tank)

and 3.5° for CC-QVCO and TS-QVCO, respectively. The proposed CC-QVCO demonstrates more robust current tolerance than TS-QVCO according to the simulation results.

4.4.6   Tolerance to Temperature

Other parameters such as chip temperature, ambient environment, and fabrication process also play important roles in a QVCO performance. Most commercial applications usually require an operational temperature range of 0–85°. Simulation results of phase noise and phase error with temperature range of -45–$125°$ is shown in Fig. 4.15. Again the CC-QVCO demonstrates better performance tolerance to temperature variations than the TS-QVCO structure.

4.5   Implementation and Measurement Results

The proposed CC-QVCO, class-C mode TS-QVCO, and SVCO have been fabricated in a 0.13-µm CMOS technology. The core areas for the three circuits are $A_{\text{CC-QVCO}} = A_{\text{TS-QVCO}} = 1.0 \times 0.35$ mm^2 and 0.35×0.5 mm^2 for SVCO, respectively. The inductors for all the implemented oscillators are of the same size with a value of

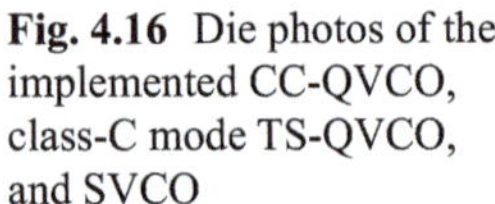

Fig. 4.16 Die photos of the implemented CC-QVCO, class-C mode TS-QVCO, and SVCO

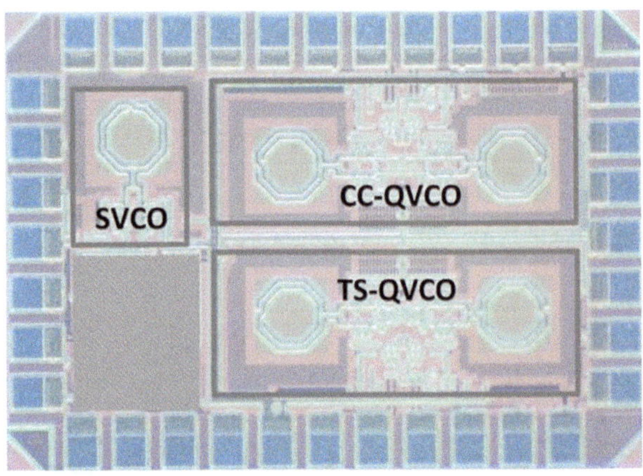

1.12 nH. Die photos for the three oscillators are shown in Fig. 4.16. Diode varactors have been used to achieve fine frequency tuning while the 2-bit MIM capacitor array are utilized for coarse tuning. The layout of the quadrature-coupling path for the proposed QVCO has been optimized to improve the matching since the mismatch in the coupling interconnections directly affects the quadrature phase error. All results were measured with a 1.2-V power supply if not specified. The phase noise and the output spectrum are measured with a spectrum analyzer E4446A

4.5.1 Upconversion Mixer and IF Baseband Signal Generation

Usually the sampling rate of an oscilloscope cannot provide adequate time resolution to measure the quadrature accuracy at 5 GHz, especially when the parasitics on the board and losses through cable will further degrade the measurement error. As a result, the measured phase accuracy using oscilloscope can be easily affected by the testing setup. One of the most popular ways for measuring the quadrature phase error is to observe the sideband rejection ratio after single-sideband upconversion mixer [7, 18]. This technique requires a quadrature IF baseband signal, which can be generated from a RC poly-phase filter. A four-stage off-chip RC poly-phase filter and an on-chip passive mixer are employed to upconvert the IF baseband signal to VCO frequency. Then, the phase error can be measured from the sideband rejection ratio at the RF spectrum.

4.5.2 Phase Noise and Frequency Range

The measured phase noise performance for the proposed CC-QVCO is shown in Fig. 4.17. Under 1.2 V supply voltage and 8.5 mA, the CC-QVCO achieves a measured phase noise of -124.04 dBc/Hz @ 1-MHz offset with center frequency of

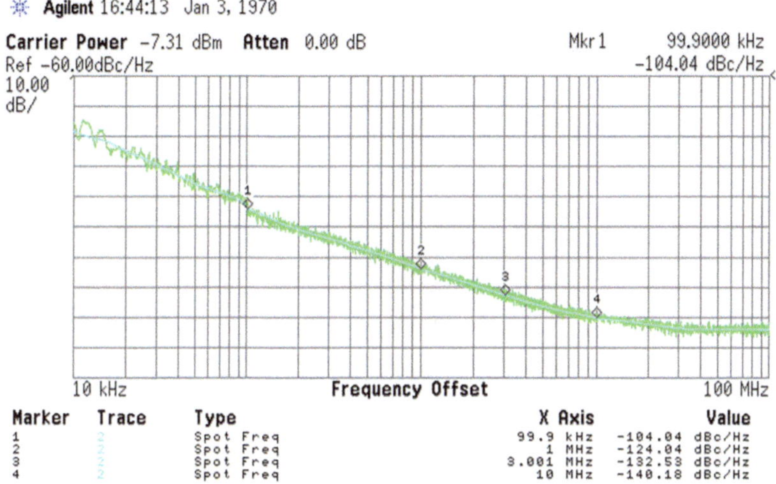

Fig. 4.17 Measured phase noise performance of the proposed CC-QVCO at 4.7 GHz with a 1.2 V power supply and 8.5 mA total current

4.7 GHz. The corresponding FoM for this phase noise performance is 187.4 dBc/Hz. Shown in Fig. 4.18 is the measured phase noise performance of the class-C mode TS-QVCO. It achieves a measured phase noise performance of -121.05 dBc/Hz @ 1-MHz offset with a center frequency of 4.9 GHz and the corresponding FoM is 184.52 dBc/Hz. The measured phase noise performance at 1-MHz offset for CC-QVCO is 3 dB better than the TS-QVCO.

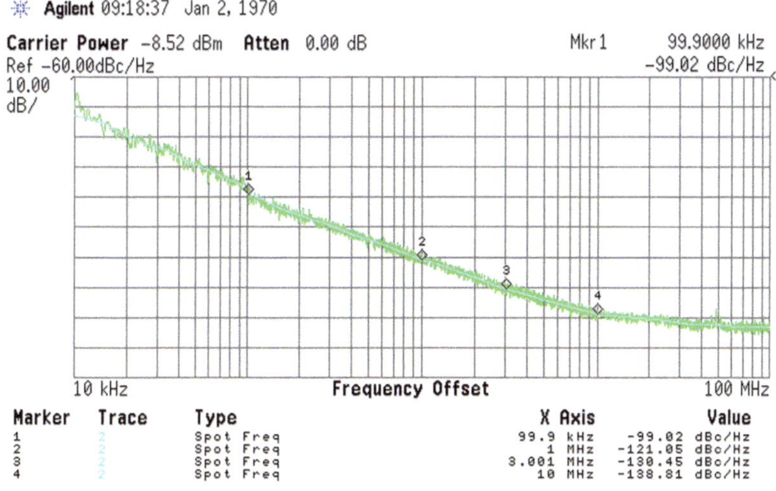

Fig. 4.18 Measured phase noise performance of the proposed TS-QVCO at 4.9 GHz with a 1.2 V power supply and 9 mA total current

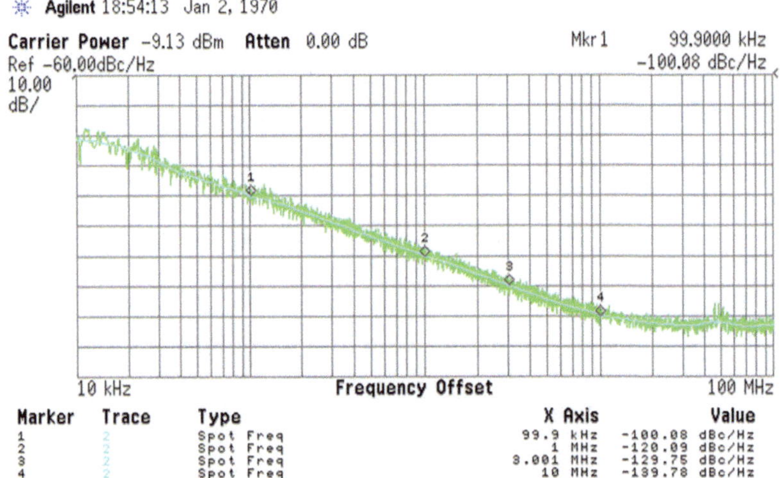

Fig. 4.19 Measured phase noise of the implemented SVCO at 5.35 GHz with a 1.2 V power supply and 4.6 mA current

The phase noise of the implemented SVCO prototype is shown in Fig. 4.19. It achieves a measured phase noise of − 120.09 dBc/Hz @ 1-MHz offset, consuming 4.6 mA from a 1.2 V supply voltage. As a result, the phase noise performance of the proposed CC-QVCO is 4 dB better than that of the implemented SVCO.

The measured frequency ranges for the implemented CC-QVCO, TS-QVCO, and SVCO are 4.3–5.27, 4.1–4.9, and 4.3–5.35 GHz, respectively. These tuning ranges correspond to frequency tuning of 21.4, 17.8, and 21.8%. The frequency tuning range of TS-QVCO is slightly lower than the other two prototypes.

4.5.3   Phase Accuracy

With the auxiliary IF signal generation circuit and upconversion mixer, the phase accuracy can be measured at the output of the upconversion mixer. The phase accuracy can be obtained from the sideband rejection (SBR) at the up-converted spectrum as shown in Fig. 4.20 for CC-QVCO. The best rejection ratio of the unwanted sideband is 60.33 dB for CC-QVCO, corresponding to 0.11° quadrature phase error, while that of the implemented TS-QVCO is 31.6 dB. Therefore, the proposed CC-QVCO achieves much better phase accuracy than the implemented TS-QVCO prototype. The relationship between phase error and image rejection is expressed as [1]

$$IR(dB) = 10log\left(\frac{4}{\theta^2}\right) \tag{4.32}$$

Figure 4.21 illustrates the measured SBR of CC-QVCO in the frequency tuning range of 4.3–5.27 GHz. The SBR ranges from 42 to 54 dB for tuning voltage below

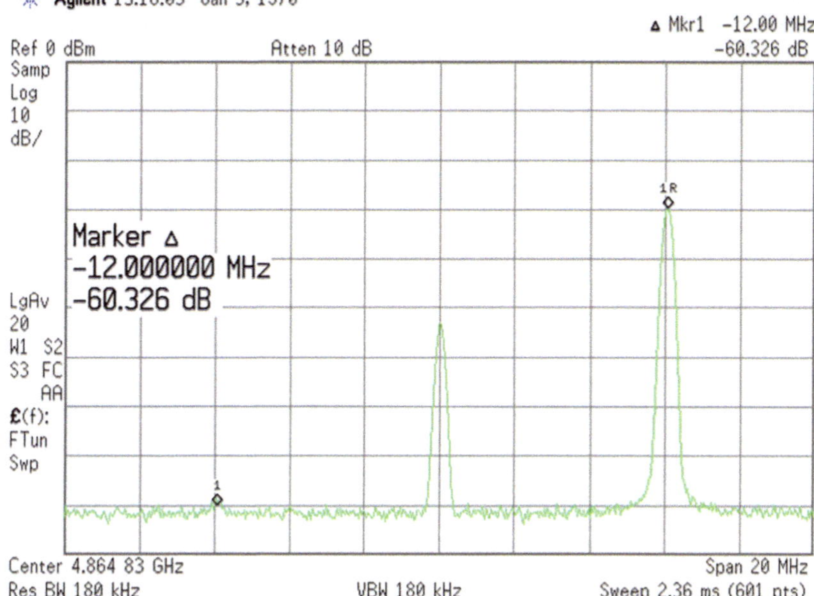

Fig. 4.20 Measured output spectrum at the output of upconversion mixer where the frequency is 4.86 GHz

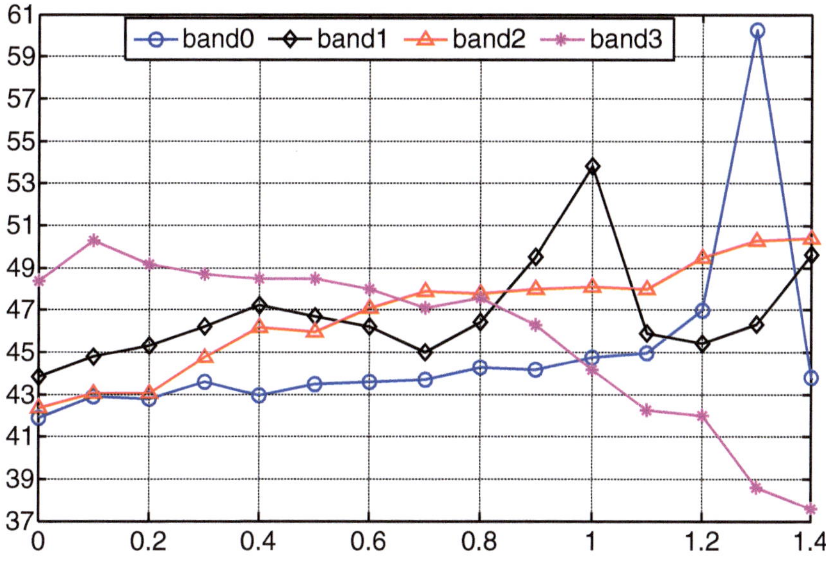

Fig. 4.21 Measured SBR of CC-QVCO across the tuning range with 1.2 V VDD

1.2 V, corresponding to phase error of $0.23°\sim0.91°$. The average SBR across the tuning range is around 45 dB. The only exception is the SBR for band 3 at tuning voltage above 1.2 V. During the testing process, we observe that the measurement result of SBR across the tuning range can be improved by more than 10 dB when the output buffers are turned off. Therefore, the practical phase error performance may be better than the measured results shown in Fig. 4.21, considering the loading effect of the buffers for phase noise measurement.

4.6   Conclusions

An NMOS QVCO with capacitive coupling and embedded leading phase delay is discussed in this chapter. The CC-QVCO achieves good phase noise performance by removing the active devices for quadrature coupling and producing phase delays needed for adjusting the timing of the coupling signals. It shows better phase noise performance than its SVCO and the implemented TS-QVCO. The problem of phase ambiguity is avoided by introducing a source degeneration capacitor. A mode rejection ratio is developed to evaluate a QVCO's capability to reject the unwanted oscillation mode. The prototype CMOS CC-QVCO was fabricated in 0.13-μm CMOS technology with measured frequency tuning range of 21.4%. The proposed CC-QVCO achieves a best quadrature phase error of $0.11°$ while consuming only 8.5 mA current from a 1.2-V power supply and occupies a core area of 1.0×0.35 mm^2. The overall performance of the three VCO prototypes is summarized in Table 4.1.

Table 4.1 Performance summary of the implemented SVCO and QVCOs

	CC-QVCO	SVCO	TS-QVCO
Technology	0.13 μm CMOS		
Power supply (V)	1.2		
Current (mA)	8.5	4.6	9.0
Frequency range (GHz)	4.3–5.27	4.3–5.35	4.1–4.9
Relatively tuning range	21.4%	21.8%	17.8%
Phase noise (dBc/Hz@1 MHz)	−124.04 (4.7 GHz)	−121.05 (4.9 GHz)	−120.09 (5.35 GHz)
Phase error	$0.23°$–$0.91°$	NC	$2.9°$
FOM (dBc/Hz)	187.4	187.24	184.52
FOM$_T$ (dBc/Hz)	199.91	199.83	196.23
Area (mm^2)	0.35	0.175	0.35

References

1. B. Razavi, RF Microelectronics Second Edition, New Jersey, USA, Prentice Hall, 2011
2. C. Liao and S. Liu, "40 Gb/s transimpedance-AGC amplifier and CDR circuit for broadband data receivers in 90 nm CMOS," IEEE J. Solid-State Circuits, vol. 43, no. 3, pp. 642–655, Mar. 2008.
3. S. L. J. Gierkink, S. Levantino, R. C. Frye, C. Samori, and V. Boccuzzi, "A low phase-noise 5-GHz CMOS quadrature VCO using superharmonic coupling," IEEE J. Solid-State Circuits, vol. 38, no. 7, pp. 1148–1154, Jul. 2003.
4. X. Li, S. shekhar, and D. J. Allstot, "Gm-boosted common-gate LNA and differential Colpitts VCO/QVCO in 0.18-µm CMOS," IEEE J. Solid-State Circuits, vol. 40, no. 12, pp. 2609–2619, Dec. 2005.
5. J. Crols and M. Steyaert, "A fully integrated 900 MHz CMOS double quadrature downconverter," in IEEE Int. Solid-State Circuits Conf. Dig. Tech. Papers, Feb. 1995, pp. 136–137.
6. P. Andreani, "A 2 GHz, 17 % tuning range quadrature CMOS VCO with high figure-of-merit and 0.6° phase error," in Proc. European Solid-State Circuits Conf., Aug. 2002, pp. 815–818.
7. A. Rofougaran, J. Rael, M. Rofougaran, and A. Abidi, "A 900 MHz CMOS LC-oscillator with quadrature outputs," in IEEE Int. Solid-State Circuits Conf. Dig. Tech. Papers, Feb. 1996, pp. 392–393.
8. A. Rofougaran, G. Chang, J. Rael, J. Chang, M. Rofougaran, P. Chang, M. Djafari, M. Ku, E. Roth, and A. Abidi, "A single-chip 900-MHz spread-spectrum wireless transceiver in 1-µm CMOS. I. architecture and transmitter design," IEEE J. Solid-State Circuits, vol. 33, no. 4, pp. 515–534, Apr. 1998.
9. P. Andreani and X. Wang, "On the phase-noise and phase-error performances of multiphase LC CMOS VCOs," IEEE J. Solid-State Circuits, vol. 39, no. 11, pp. 1883–1893, Nov. 2004.
10. A. W. L. Ng and H. C. Luong, "A 1-V 17-GHz 5-mW CMOS quadrature VCO based on transformer coupling," IEEE J. Solid-State Circuits, vol. 42, no. 9, pp. 1933–1941, Sep. 2007.
11. A. Mirzaei, M. E. Heidari, R. Bagheri, S. Chehrazi, and A. A. Abidi, "The quadrature LC oscillator: a complete portrait based on injection locking," IEEE J. Solid-State Circuits, vol. 42, no. 9, pp. 1916–1932, Sep. 2007.
12. F. Zhao and F. F. Dai, "A 0.6-V quadrature VCO with enhanced swing and optimized capacitive coupling for phase noise reduction," IEEE Trans. Circuits Syst. I, Reg. Papers, vol. 59, no. 8, pp. 1694–1705, Aug. 2012.
13. J. van der Tang, P. van de Ven, D. Kasperkoviz, and A. van Roermund, "Analysis and design of an optimally coupled 5-GHz quadrature LC oscillator," IEEE J. Solid-State Circuits, vol. 37, no. 5, pp. 657–661, May 2002
14. F. Zhao and F. F. Dai, "A Capacitive-Coupling Technique with Phase Noise and Phase Error Reduction for Multi-Phase Clock Generation," in Proc. IEEE Custom Integrated Circuits Conf., Sep. 2014, pp. 1–4.
15. M. Danesh and J. R. Long, "Differentially driven symmetric microstrip inductors," IEEE Trans. Microw. Thoery Tech., vol. 50, no. 1, pp. 332–341, Jan. 2002.
16. A. Hajimiri and T. H. Lee, "Design issues in CMOS differential LC oscillators," IEEE J. Solid-State Circuits, vol. 34, no. 5, pp. 896–909, May. 1999.
17. D. B. Leeson, "A simple model of feedback oscillator noises spectrum," Proc. IEEE, vol. 54, pp. 329–330, Feb. 1966.
18. P. Andreani, A. Bonfanti, L. Romano, and C. Samori, "Analysis and design of a 1.8-GHz CMOS LC quadrature VCO," IEEE J. Solid-State Circuits, vol. 37, no. 12, pp. 1737–1747, Dec. 2002.
19. A. Mazzanti and P. Andreani, "Class-C harmonic CMOS VCOs, with a general result on phase noise," IEEE J. Solid-State Circuits, vol. 43, no. 12, pp. 2716–2729, Dec. 2008.

Chapter 5
Design and Analysis of A Low Power QVCO with Capacitive-Coupling Technique

5.1 Background

Phase noise and phase accuracy are two essential specifications for quadrature signal generation since the two aspects directly affect the quality of the received or transmitted signal in a wireless communication system. The ever-growing demand for chip-level integration of multiband transceiver continues imposing tighter phase noise performance specifications for radio-frequency (RF) carrier generation. Quadrature signals with phase accuracy and no phase ambiguity are critical for image-rejection transceivers since they directly affect the polarity and the outcome of the complex mixers. Phase error presented in the quadrature signals will add to the error of a baseband signal and deteriorate the bit error rate (BER) or of a communication system. Thus, a high performance quadrature signal generation technique with both low noise and decent phase accuracy is highly desirable for complex signal modulation and demodulation.

To eliminate noise degradation introduced by the coupling mechanism, noiseless components such as transformer, inductor, and capacitor can be used for coupling. A QVCO with transformer coupling which is based on the technique of superharmonic coupling [1] shows good phase noise performance with the expense of inductor area. An energy-circulating QVCO with inductive coupling can achieve better phase noise performance than the single-phase VCO of the same kind [2], yet it comes at the cost of additional area of two inductors. In order to reduce the area of a coupling transformer, the secondary coupling tank can share the tank area with the resonant tank and it can achieve a decent figure-of-merit (FoM) [3]. However, transformer models are either not accurate or not available in most commercial CMOS technology and it requires extra effort to develop an accurate transformer model. Therefore, QVCO with capacitive coupling techniques [4–6] have been developed to simplify the circuit design with good noise performance and small area.

F. Zhao, F. F. Dai, *Low-Noise Low-Power Design for Phase-Locked Loops,*
DOI 10.1007/978-3-319-12200-7_5

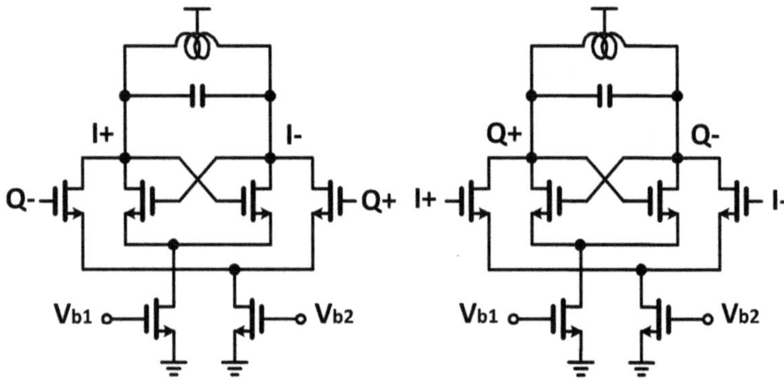

Fig. 5.1 Conventional quadrature VCO with parallel coupling transistors

Various QVCO coupling mechanisms have been developed in search of improved phase noise performance, yet another important aspect of the QVCO design, the phase ambiguity, is often overlooked. The understanding of the phase ambiguity and the stability is critical since a typical QVCO may operate at either one of its two stable modes with different phase relationships. Each stable mode corresponds to $+90°$ or $-90°$ phase relationship between the two outputs of the QVCO. However, quadrature signals with deterministic phase relationship are often required for proper image rejection in RF receivers [7]. The phenomenon of the bimodal oscillation has been observed and phase shifter in the coupling path can help solving this problem [8–10]. Theoretical analysis and experimental results prove that the phase shift of 90° introduced in the quadrature-coupling path provides optimum phase noise performance and minimum phase error arising from mismatch between two VCO cores [10]. However, the phase shift of 90° has to be implemented with poly-phase shifters [10], additional active devices stages [11], or source degenerated phase shifter [12] for QVCO using parallel coupling transistors.

For a conventional QVCO with parallel coupling transistors as shown in Fig. 5.1, the iVCO and the qVCO couple with each other at the gate of the coupling transistors and the largest energy injection happens at the zero-crossings of the VCO output swing. According to the impulse sensitivity function (ISF) theory [13], the VCO phase noise is most sensitive to disturbance near the zero-crossings of the oscillation. Consequently, the phase noise of the quadrature outputs is degraded due to the fact that the amplitude-to-phase noise conversion in this topology is largest at their zero-crossings. For this reason, a QVCO with parallel coupling ends up with worse phase noise than that of its single-phase counterpart.

The current trend of technology scaling presents challenges for circuit designs. Feature size shrinking forces the power supply drop below 1 V. Lowered supply voltage limits the output swing that can be generated, which further limits the phase noise that an oscillator can achieve. A Colpitts QVCO with enhanced swing and capacitive coupling technique [14] for low phase noise performance has been proposed for a 0.6-V supply voltage. The capacitive coupling (CC)-QVCO, as shown in Fig. 5.2, not only achieves low-phase noise performance under a low supply

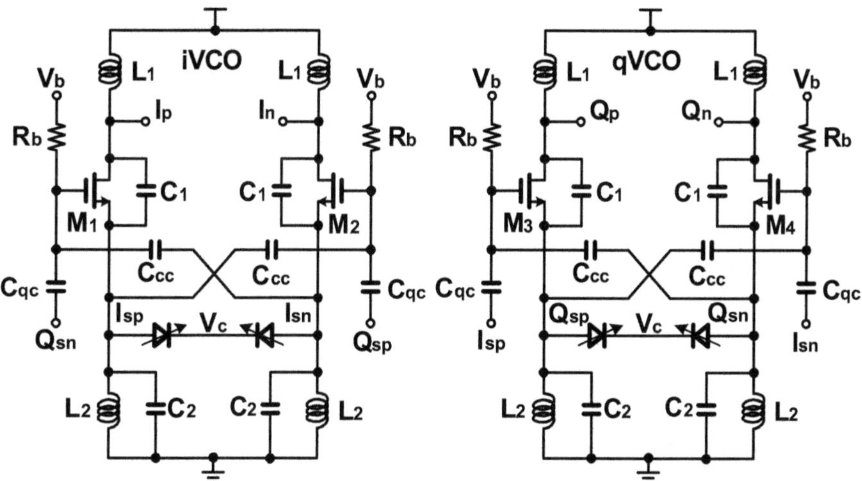

Fig. 5.2 Proposed QVCO with optimized capacitive coupling and intrinsic phase shift

voltage, but also guarantees stable oscillation with an intrinsic phase shift in the coupling path.

This chapter presents the details of the proposed capacitive-coupled QVCO (CC-QVCO) [14]. In Sect. 5.2, the architecture of the CC-QVCO, noise-reduction technique, and optimization of the capacitive coupling are introduced. Moreover, the transconductance- (effective G_m-) enhancement technique for power reduction and the intrinsic phase shift for stable oscillation is analyzed in Sect. 5.2. Section 5.3 provides the implementation and experimental results for the proposed CC-QVCO. Finally, conclusions are drawn in Sect. 5.4.

5.2   CC-QVCO with Noise Reduction and Stable Oscillation

5.2.1   Architecture of the CC-QVCO

As shown in Fig. 5.2, instead of using noisy transistors for quadrature signal coupling, capacitive coupling is employed to improve the phase noise performance of the QVCO. To achieve large output swing required for good phase noise performance under a low supply voltage around 0.6 V, an enhance-swing (ES) Colpitts VCO structure similar to the one described in [15] is adopted. Different from simple ES Colpitts VCO, G_m- enhancement technique is employed using the cross-coupled capacitors C_{cc} to reduce the power consumption. The proposed CC-QVCO is composed of two such G_m-enhanced VCO cores and four quadrature-coupling capacitors C_{qc}. The coupling-strength factor m between the iVCO and qVCO is defined as

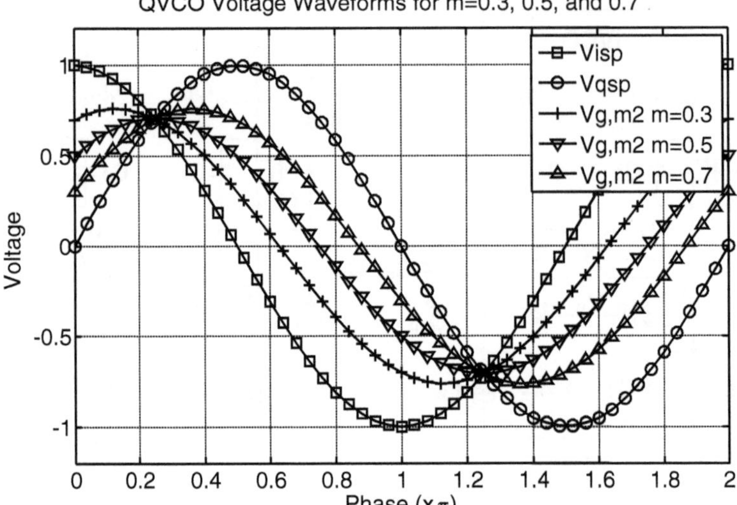

Fig. 5.3 Voltage waveforms for different coupling-strength factor m

$$m = \frac{C_{qc}}{C_{cc} + C_{qc}} \tag{5.1}$$

Assuming the transient voltages of the quadrature output signals as $V_{isp} = V_0\cos(\omega_0 t)$ and $V_{qsp} = V_0\sin(\omega_0 t)$, the voltage signal at the gate of M_2 is

$$V_{g,M2}(t) = mV_{qsp}(t) + (1-m)V_{isp}(t) \tag{5.2}$$

The voltage waveforms with different coupling-strength factor are illustrated in Fig. 5.3. As it can be seen from the figure, the smaller the coupling strength factor m is, the farther the peak of $V_{g,M2}$ deviates from the zero-crossing of V_{isp} and V_{qsp}. As the voltage at the drain of each transistor has the same phase as its source voltage, the voltage peak on the gates can also be shifted away from the zero-crossings of the output voltage. As a result, the amplitude of the gate voltage is no longer the maximum during the zero-crossings of the VCO output swing. Moreover, the effective ISF for the CC-QVCO can also be improved. Therefore, the amplitude-to-phase noise conversion between the two VCO cores is reduced and the phase noise performance of the CC-QVCO is improved.

In addition, phase noise is further improved by placing diode junction varactors with reference to the ground. The quality factor reduction caused by the parasitic diodes has been avoided because the VCO tank on n-type anode has been isolated from substrate since the p-type cathode is connected to a DC bias voltage [16, 17]. The combination of these techniques described above enables the proposed CC-QVCO with low phase noise (− 122 dBc/Hz @ 1-MHz offset) and low power consumption (4.2 mW).

5.2.2   *Colpitts VCO Core with G_m-Enhancement*

A Colpitts VCO features superior phase noise characteristics than cross-coupled VCO since the noise injection from active devices for the former structure is at the minimum of the tank voltage when the ISF is low [13, 18]. Unfortunately, a Colpitts VCO requires large transconductance which means more power to meet the start-up conditions in the presence of process-voltage-temperature (PVT) variations. Therefore, high power dissipation is necessary to ensure reliable start-up.

Figure 5.4 shows the half circuits of different Colpitts VCO topologies. The derivation of the small-signal admittance for Colpitts VCO with current tail as shown in Fig. 5.4a is straightforward and can be written as

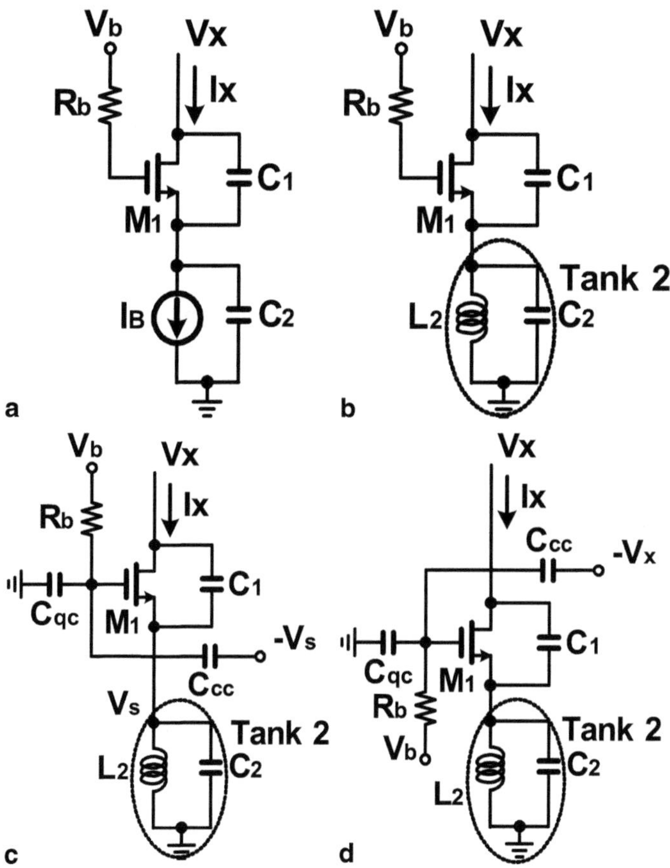

Fig. 5.4 Half circuits of differential Colpitts VCOs used to analyze the start-up condition and resonance frequency. **a** conventional structure with current tail. **b** ES VCO. **c** ES VCO with cross-coupled positive feedback at source. **d** ES VCO with cross-coupled positive feedback at drain

$$Y_{in,\text{Itail}} = \frac{s^2 C_1 C_2}{g_m + s(C_1 + C_2)} \tag{5.3}$$

where g_m is the transconductance of M_1. The admittance of an ES-Colpitts VCO shown in Fig. 5.4b with tank 2 to enhance the signal swing is given by

$$Y_{in,ES} = \frac{sC_1}{1 + g_m Z_{L2} + sC_1 Z_{L2}} = \frac{sC_1\left(1 + s^2 L_2 C_2\right)}{1 + s^2 L_2 \left(C_1 + C_2\right) + g_m s L_2} \tag{5.4}$$

$$Z_{L2} = sL_2 \,\|\, \frac{1}{sC_2} \,\|\, R_{P2} \tag{5.5}$$

Equation (5.4) is based on ideal lossless inductor L_2, i.e., $R_{p2} = \infty$. Shown in Fig. 5.4c and d are other two Colpitts VCO structures with G_m-enhancement. ES VCO of Fig. 5.4c places the cross-coupled capacitor at the source. The admittance looking into the half-circuit can be derived with the Kirchhoff's circuit laws (KCL). The voltage at the drain can be expressed as

$$I_X \frac{1}{sC_1} + I_X g_m (2 - m) Z_{L2} \frac{1}{sC_1} + I_X Z_{L2} = V_X \tag{5.6}$$

The admittance for the ES VCO with G_m-enhancement placed at source (ESEG$_m$-S) is defined as

$$Y_{in,ESEGm-S} = \frac{I_X}{V_X} = \frac{sC_1}{1 + (2 - m)g_m Z_{L2} + sC_1 Z_{L2}} \tag{5.7}$$

where g_m is the transconductance of M_1. By assuming an ideal lossless inductor L_2, the admittance for ESEG$_m$-S VCO can be rewritten as

$$Y_{in,ESEGm-S} = \frac{sC_1(1 + s^2 L_2 C_2)}{1 + s^2 L_2 (C_1 + C_2) + (2 - m)g_m s L_2} \tag{5.8}$$

Similarly, the admittance for ESEG$_m$-D VCO as shown in Fig. 5.4d can be derived as

$$Y_{in,ESEGm-D} = \frac{sC_1 - (1 - m)g_m}{1 + (g_m + sC_1)Z_2} = \frac{\left[sC_1 - (1 - m)g_m\right]\left(1 + s^2 L_2 C_2\right)}{1 + s^2 L_2 \left(C_1 + C_2\right) + g_m s L_2} \tag{5.9}$$

The real parts of those equations represent the negative transconductance required to start the oscillator. The larger the absolute value of the transconductance is, the smaller is the required power consumption for start-up. Oscillators will fail to start oscillation when the negative admittance cannot compensate the tank loss. The real admittances for the four VCO topologies are expressed as follows:

$$Re\left[Y_{\mathrm{in,Itail}}\right] = \frac{-g_{\mathrm{m}}\omega^2 C_1 C_2}{g_{\mathrm{m}}^2 + \omega^2 (C_1 + C_2)} \qquad (5.10)$$

$$Re\left[Y_{\mathrm{in,ES}}\right] = \frac{-g_{\mathrm{m}}\omega^2 L_2 C_1 \left(\omega^2 L_2 C_2 - 1\right)}{\left[1 - \omega^2 L_2 (C_1 + C_2)\right]^2 + \left(g_{\mathrm{m}}\omega L_2\right)^2} \qquad (5.11)$$

$$Re\left[Y_{\mathrm{in,ESEGm\text{-}S}}\right] = \frac{-(2-m)g_{\mathrm{m}}\omega^2 L_2 C_1 \left(\omega^2 L_2 C_2 - 1\right)}{\left[1 - \omega^2 L_2 (C_1 + C_2)\right]^2 + \left[(2-m)g_{\mathrm{m}}\omega L_2\right]^2} \qquad (5.12)$$

$$Re\left[Y_{\mathrm{in,ESEGm\text{-}D}}\right] = -g_{\mathrm{m}}\left(\omega^2 L_2 C_2 - 1\right) \times \frac{\omega^2 L_2 C_1 - (1-m)\left[1 - \omega^2 L_2 (C_1 + C_2)\right]}{\left[1 - \omega^2 L_2 (C_1 + C_2)\right]^2 + \left(g_{\mathrm{m}}\omega L_2\right)^2} \qquad (5.13)$$

Figure 5.5a shows the calculated real admittances of the four VCO structures. As shown in the frequency range of $5 \sim 6$ GHz, the conductance of $ESEG_{\mathrm{m}}$-S VCO is about 1.5 times that of ES VCO and thus relaxes the start-up requirement. Compared with conventional Colpitts VCO with ideal current tail, the improvement at $f = 5.5$ GHz is about 35 %. Therefore, the power consumption is reduced and improved FoM can be achieved. The improvement has been verified through simulation and the simulated admittances are shown in Fig. 5.6. Although the simulated conductance improvement is smaller than the calculation result, the Colpitts VCOs with G_{m}- enhancement as shown in Fig. 5.4c and d still achieve lower power consumption than the other two structures. The discrepancies between the Fig. 5.5 and Fig. 5.6 are caused not only by using simplified small-signal transistor models with first-order approximation, but also by neglecting C_{qc}, C_{cc}, and other parasitic capacitances for deriving the analytic expressions. However, Fig. 5.5 gives first-order approximation of the admittances. The magnitude of negative G_{m} decreases when frequency is reduced, i.e., it becomes more difficult for the VCOs to meet the start-up condition as frequency decreases. After a certain frequency value, the G_{m} becomes positive and peaks at the resonant frequency of tank 2 as shown in Fig. 5.5a and Fig. 5.6a. The resonant frequency of tank 2 should be placed far below the VCO resonance frequency to maintain a sufficient margin for stable oscillation.

The resonance frequency of the Colpitts VCO core is determined by the inductor L_1 and the equivalent capacitance looking into the drain terminal. The equivalent capacitor without considering parasitic capacitances can be obtained from the imaginary part of Eqs. (5.3), (5.4), (5.8), and (5.9), as shown in Fig. 5.5b, while the simulation result for the equivalent capacitor is shown in Fig. 5.6b. The simulated

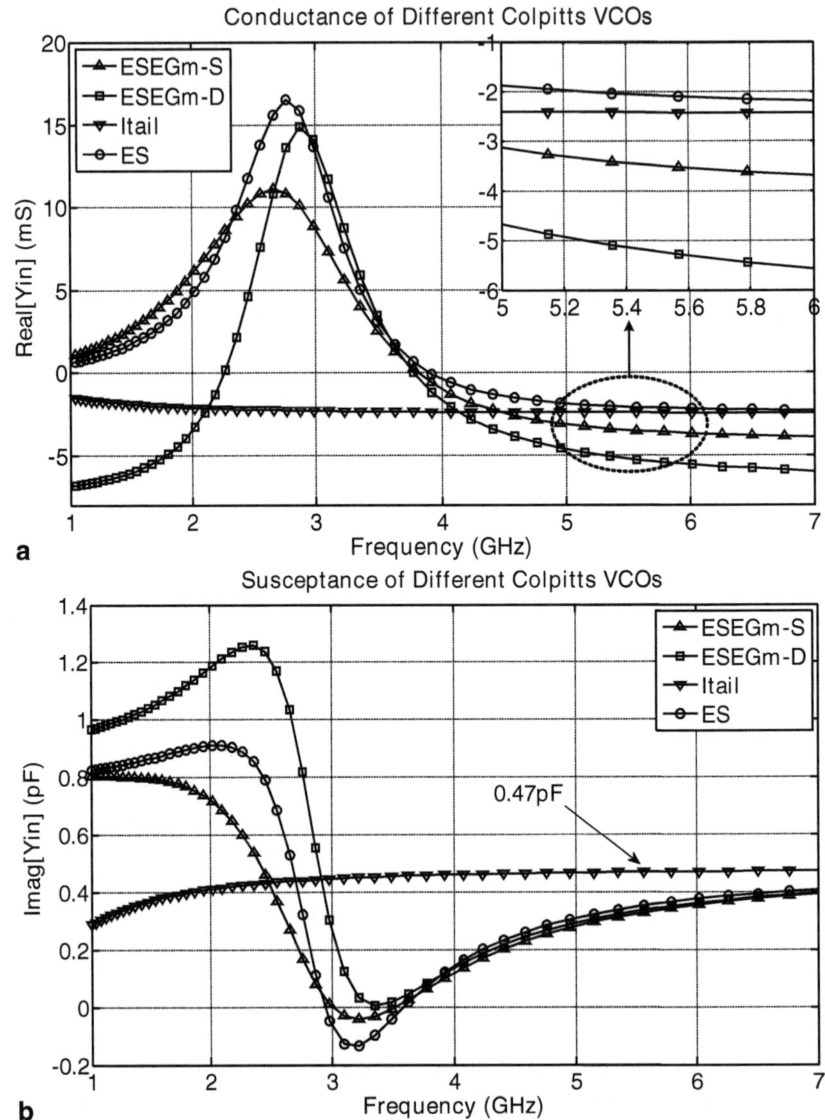

Fig. 5.5 Calculation results of **a** conductance, and **b** susceptance for different Colpitts VCOs. Component values used for calculation are as follows: $C_1 = 0.8$ pF, $C_2 = 1.2$ pF, $L_2 = 1.25$ nH, $g_m = 10.3$ mS, $Q_{L2} = 15$

equivalent capacitor for $ESEG_m$-D VCO is larger than the other two VCOs with bottom inductors because of the directly added quadrature-coupling capacitors. For the two Colpitts VCO structures shown in Fig. 5.4b and c, which have inductors at source terminals, the equivalent capacitance looking into the drain is reduced since inductor L_2 cancels part of the capacitor at the cost of the bottom inductor. However, the primary goal of the bottom inductor in this design is to enhance the

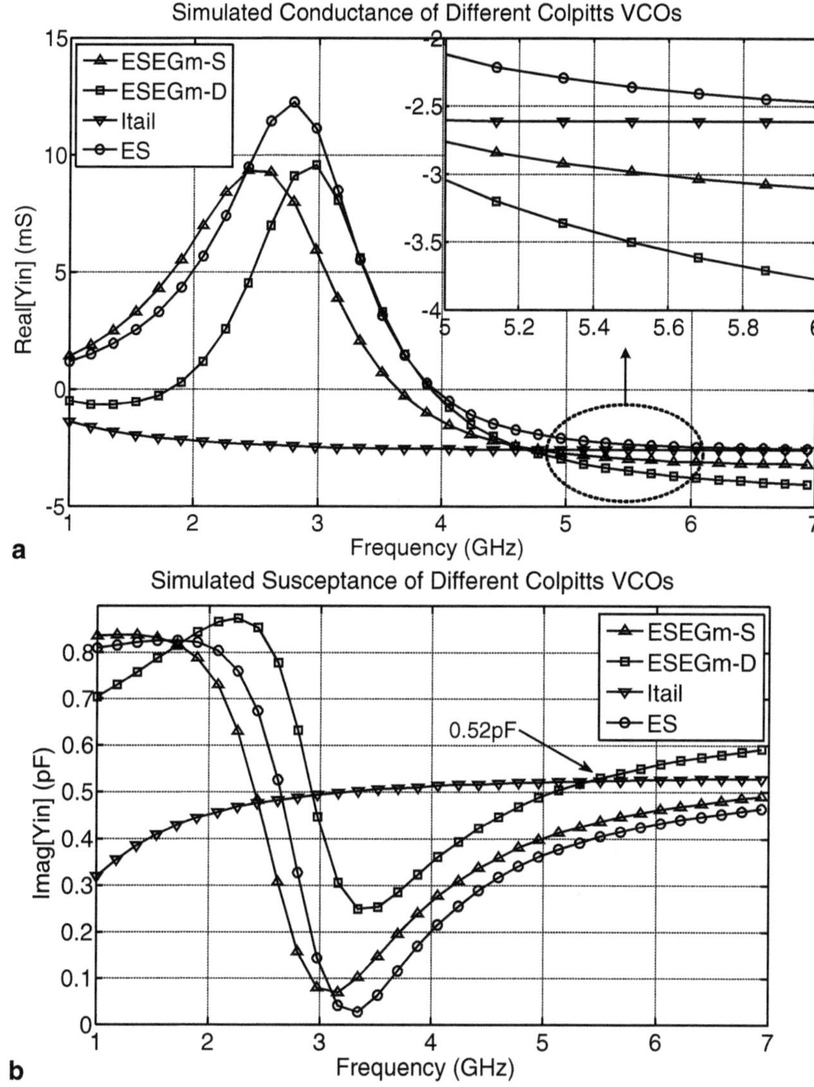

Fig. 5.6 Simulation results of **a** conductance, and **b** susceptance for different Colpitts VCOs. Components used for simulation are the same as calculation

signal swing under a low supply voltage. As a result, the resonance frequency is increased compared with a conventional Colpitts VCO. This feature is very useful for RF frequency VCO design since the parasitic capacitance starts to dominate at high frequency.

Although the conductance of $ESEG_m$-D VCO can save much more power than that of $ESEG_m$-S VCO, the latter structure is used because its performance is less sensitive to the mismatches produced by the quadrature-coupling path than the former. This can be understood by observing the Colpitts VCO structure giv-

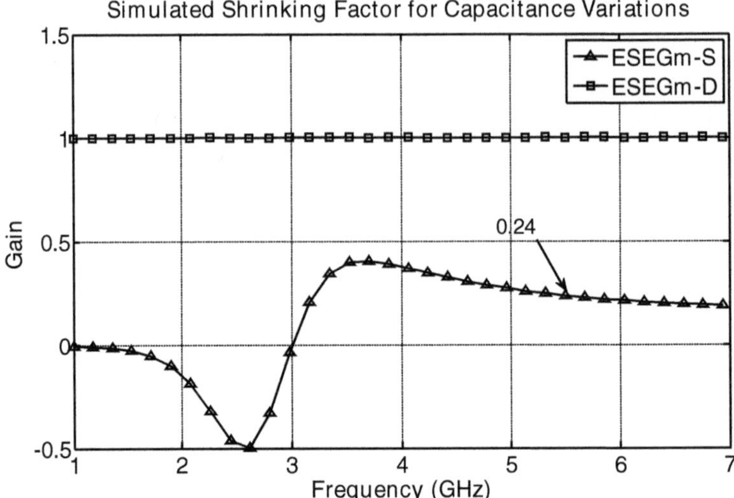

Fig. 5.7 Simulation results of the shrinking factors for ESEGm-D VCO and ESEGm-S VCO

en in Fig. 5.4a. The equivalent capacitance at the drain can be approximated as $C_{tot} = C_{drain} + C_1 \parallel C_2$ around resonant frequency. The capacitance variation ΔC_2 at the source is shrunk by a factor of $n^2 = C_1^2 / (C_1 + C_2)^2$ and n is usually smaller than 0.4 for good phase noise performance. However, the capacitance variation ΔC_{drain} at the drain directly adds to the total capacitance. Figure 5.7 shows the shrinking factor of capacitance variations for $ESEG_m$-S VCO and $ESEG_m$-D VCO. The capacitance variations are applied to the source for $ESEG_m$-S VCO and the drain for $ESEG_m$-D VCO. From the Fig. 5.7, it is known that the shrinking factor for the $ESEG_m$-S VCO is about one fourth of that of the $ESEG_m$-D VCO around the target frequency. Therefore, the $ESEG_m$-S VCO suffers less from the mismatch in the quadrature-coupling path than the $ESEG_m$-D VCO.

5.2.3   Noise Reduction for the CC-QVCO

Ideally the phase noise of a QVCO can be reduced by 3 dB compared to a single-phase VCO that draw half of the current of the QVCO, and a phase noise normalized to the power consumption would be the same as its single phase counterpart [19]. This assumption does not take into account various effects that have impact on the phase noise performance, such as additional noise generated by the coupling devices and the reduction of effective quality factor of the LC tanks. On the other hand, the coupled signal is usually at its maximum when the QVCO is most susceptible to noise, i.e., when the two VCO cores inject noise to each other during the zero-crossing point of their output swings. Due to both the additional noise introduced by coupling transistors and noise injection around the most sensitive time of output signals, the phase noise normalized to power consumption of the QVCO

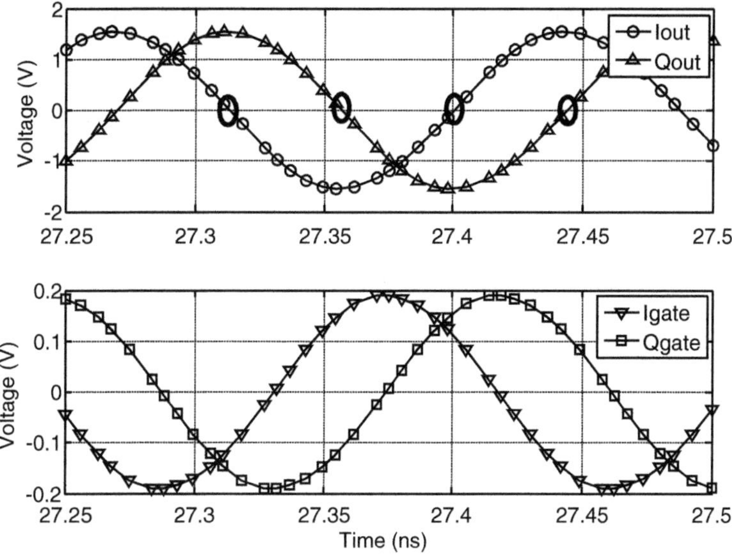

Fig. 5.8 Simulation results of CC-QVCO outputs and coupling signals with $m=0.4$. The phase difference between the zero-crossing of Iout or Qout and the peak of Igate or Qgate is about 55°

based on series or parallel coupling [19–22] can only be close, but not as good as that of its single-phase counterpart.

In order to lower the phase noise, it is beneficial to reduce the voltage swings of the coupled signals at their zero-crossing time. By using cross-coupled capacitors for G_m-enhancement and quadrature-coupling capacitors for quadrature generation, the voltages on the gate can be shaped for better noise performance. Figure 5.8 shows the simulated transient voltages for the CC-QVCO with $m=0.4$. It is obvious that the voltage maxima of the coupling signal have been shifted away from the zero-crossings of the output signals. As a result, the amplitude-to-phase noise conversion between the two VCO cores is reduced and the phase noise performance of the CC-QVCO is improved beyond what can be achieved by its single-phase counterpart.

In order to verify the noise improvement, the ISF and effective ISF (ISF_{eff}) of the QVCO and SVCO are obtained using the direct impulse response measurement method of [13] implemented in MMSIM 10.2. The QVCO and SVCO are simulated using the same tank inductance and are tuned to oscillate at a center frequency of 5.8 GHz. The QVCO including two VCO cores draws twice the current of the SVCO. Figure 5.9 shows the simulated ISFs of the QVCO versus that of the single-phase VCO (SVCO). The noise-modulating function (NMF) is defined as the instantaneous drain current divided by the peak drain current over an output signal cycle. The ISF_{eff} is defined as the product of ISF and NMF. The simulation result shows that the proposed CC-QVCO achieves lower ISF and effective ISF_{eff} than its SVCO core. The corresponding coefficient ratio of c_{0svco} to c_{0qvco} is about

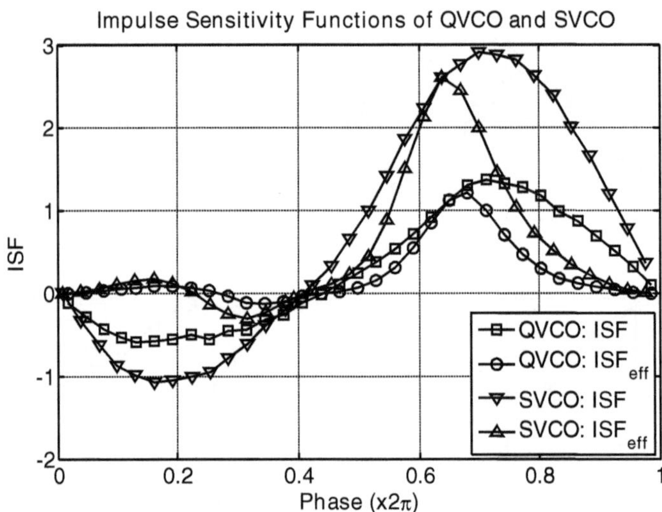

Fig. 5.9 ISF and $\mathrm{ISF}_{\mathrm{eff}}$ for CC-QVCO and SVCO with $m=0.4$, respectively

2.1, which means the noise power resulted from the transistors used in CC-QVCO will be improved by 6.4 dB.

Figure 5.10 shows the phase noise simulation results of the CC-QVCO and its SVCO core. The proposed CC-QVCO achieves 3.6~5.2-dB phase noise improve-

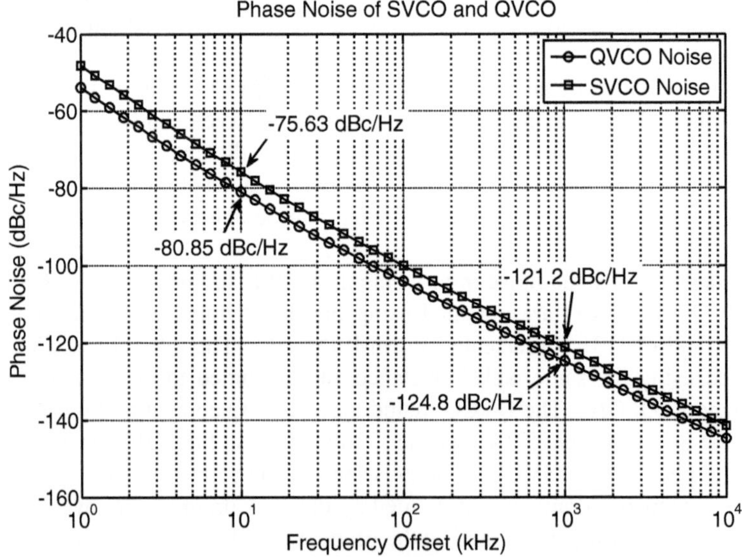

Fig. 5.10 Simulation results of phase noise for SVCO and CC-QVCO with $m=0.4$

ment at the frequency offset of 10 kHz to 1 MHz when compared with its SVCO of the same kind. The noise summary for 1-MHz offset shows that each of the four transistors for the CC-QVCO contributes a noise power of $4.90 \times 10^{-14}\,V^2/Hz$, while each of the two transistors contributes a noise power of $1.92 \times 10^{-13}\,V^2/Hz$ for the SVCO. Therefore, the noise improvement resulted from the transistors can be approximated as

$$c_{0,\text{improve}}\,(\text{dB}) = 10\log\frac{1.92 \times 10^{-13}}{4.90 \times 10^{-14}} = 5.93\text{dB} \tag{5.14}$$

This value is very close to the simulated ISF_{eff} improvement of 6.4 dB. It proves that the CC-QVCO can achieve better phase noise performance than its SVCO core because of the reduced ISF_{eff}. Since the capacitive coupling does not use devices that introduce extra noise, the CC-QVCO accomplishes 3-dB phase noise improvement predicted by the theory under ideal condition [19]. The additional noise improvement beyond 3 dB for the proposed CC-QVCO is caused by the reduced ISF_{eff} as shown in Fig. 5.7. At lower frequency offset, the noise improvement becomes more obvious than that obtained at high frequency offset, since the flicker noise of transistors dominates the overall noise performance at low frequency offset and the improvement can be more than 3 dB.

It is for the mechanism described above, that the proposed CC-QVCO outperforms the most of QVCOs published so far with good phase noise, low power consumption, and small area.

5.2.4   Optimization of Capacitive Coupling

In this section, the optimization of capacitive coupling strength factor m as defined in Eq. (5.1) is discussed. The phase noise improvement of CC-QVCO compared with its SVCO counterpart depends on the coupling-strength factor m. On the other hand, the coupling strength should be as large as possible to reduce the phase error. Thus, the selection of m is a tradeoff between noise improvement and phase error. Regardless of the tradeoff, the proposed CC-QVCO is advantageous over conventional QVCO structure for the following two reasons: (i) it completely eliminates the noise sources associated with the transistors used for quadrature coupling; (ii) it provides phase noise improvement beyond 3-dB theoretical prediction; especially the flicker noise can be improved further because of the reduced ISF_{eff}.

Shown in Fig. 5.11 are the simulated phase noise improvement and the phase errors for different coupling strength factor m. The simulation for both the phase noise and the phase error is based on the assumption of 1 % mismatch for the LC tank. It can be seen that the phase noise improvement is relatively constant around 3.5 dB @ 1-MHz offset when m is between 0.3 and 0.5. The phase noise improvement for lower offsets reaches their peaks when m is around 0.25. The phase noise improvement gradually disappears as m approaches 1. When m is equal to 0, i.e.,

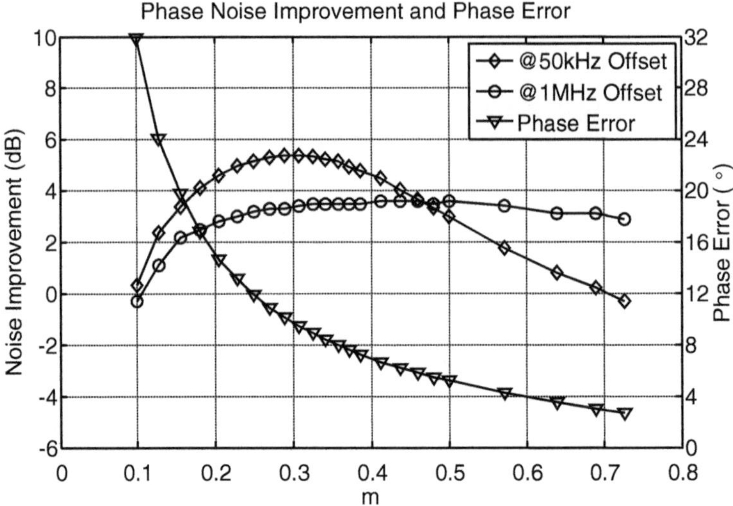

Fig. 5.11 Simulation results of phase noise improvement and phase error for different coupling strength factor m with $C_{\text{tankq}} = 1.01 C_{\text{tanki}}$

quadrature coupling disappears, the two VCO cores become independent to each other and hence fail to produce quadrature outputs. With the mismatch included in the LC tank, the noise improvement drops to 0 when m becomes 0.1 instead of 0. Given the 1 % tank mismatch, the two VCO cores are relatively independent from each other and cannot produce quadrature outputs when m is smaller than 0.1. On the other hand, the phase error increases rapidly as m approaches 0. Therefore, there is an optimum point of m to achieve the best phase noise improvement with acceptable phase error. Furthermore, the VCO design cares more about the out-of-band noise at large offset frequency since the close-in noise can be filtered by the phase-locked-loop (PLL). Considering all the factors described above, the coupling strength factor of 0.4 is chosen to implement the proposed CC-QVCO.

5.2.5 Intrinsic Phase Shift to Avoid Phase Ambiguity

This section starts with an introduction to the linear QVCO model followed by derivations for the intrinsic phase shift of the proposed CC-QVCO.

To generate quadrature outputs, two stand-alone, nominally identical VCO cores have to be coupled with each other with active devices or passive components. However, the oscillation frequency of a QVCO will depart from the resonance frequency of individual VCO cores because of the coupling mechanisms. The behavior of the two VCO cores including coupling effects can be modeled by a simplified linear model [9, 11, 19], as shown in Fig. 5.12a, where $G_{\text{m}}(s)$ provides a negative resistance compensating the energy loss due to the equivalent resistance R of an

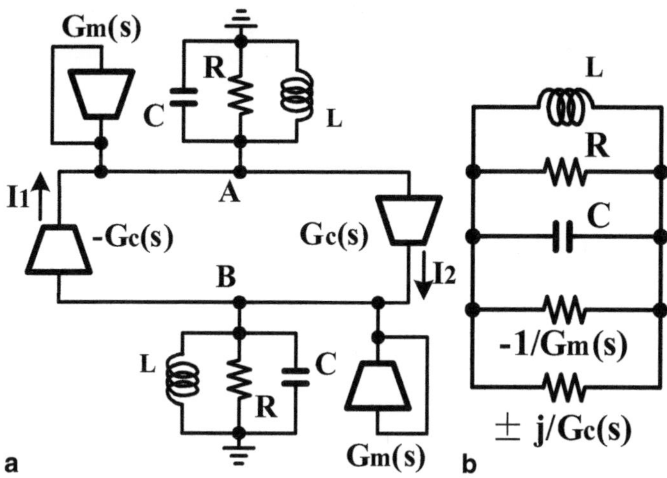

Fig. 5.12 **a** Linear model of quadrature oscillator. **b** Equivalent model of individual VCO with coupling effects

LC tank, and $G_c(s)$ represents the transconductance of the coupling mechanism between the two VCO cores.

Figure 5.12b presents the equivalent model for each VCO with quadrature-coupling mechanism. Since one VCO core may lead or lag the other VCO core by 90° phase, $\pm j$ is introduced to the coupling transconductance. For a conventional QVCO with parallel coupling transistors [23], the frequency deviation can be found with steady state analysis as

$$\Delta\omega = \omega_0 \pm \frac{G_c(j\omega_0)}{2C} \tag{5.15}$$

where the positive or negative sign depends on whether the iVCO leads or lags the qVCO by 90°; ω_0 is the resonance frequency of individual LC tank; C is the capacitor of the LC tank; and G_c is the equivalent transconductance of the quadrature-coupling mechanism between the two VCO cores. From Eq. (5.15), it is obvious that the larger the G_c is, the larger the frequency deviation from ω_0 becomes, which worsens the quality factor of the LC tank. It is desirable to decrease the coupling strength to improve the phase noise performance. The tradeoff between the phase noise and phase error calls for a solution that can maintain the phase noise performance without sacrificing the phase error.

Another commonly seen problem for a QVCO design is the phase ambiguity, i.e., the phase relations between the QVCO outputs could be either +90° or −90°. It is essential to provide 90°-quadrature phase signals with deterministic phase relationship since a receiver or transmitter which has been hard-wired to the QVCO outputs cannot distinguish complex signals if the output phases are ambiguous, i.e., the wanted sideband might be suppressed and instead the image signal might be detected after the image rejection receiver. Although the asymmetry between two VCO cores can help the QVCO to operate in one of the two stable modes, the bimodal oscillation can still exist due to PVT variations. Usually, a phase shifter could be introduced in the quadrature-coupling path to allow only one deterministic stable quadrature outputs, either $+90°$ or $-90°$ [8–12]. To allow only one modal oscillation, the phase shift can be introduced by using cascode transistor [8]. From theoretical point of view, $90°$-phase shift in the coupling path achieves not only the minimum phase noise performance, but also the best tolerance to component mismatches between the two VCO cores [10, 11]. However, those coupling mechanism with phase shifter still deteriorate the phase noise performance because of the extra noise from the coupling transistors.

The noise degradation and phase ambiguity between the two VCO cores can be solved simultaneously with the proposed Colpitts CC-QVCO. The CC-QVCO has an intrinsic phase shift in the quadrature-coupling path and it can avoid the problem associated with the bimodal oscillation. To illustrate this effect, a linear model similar to the one shown in Fig. 5.12a needs to be developed. Unlike a conventional QVCO with parallel coupling, the proposed CC-QVCO is based on Colpitts VCO structure and the quadrature-coupling transconductance $G_c(s)$ is not intuitive.

According to the linear model of Fig. 5.12a, the $G_c(s)$ can be found by grounding node B and applying a voltage source ΔV at node A. The $G_c(s)$ can be defined as the ratio of the current I_2 flowing into ground to the voltage source applied at node A. Similar method can be used to find the quadrature-coupling transconductance for the CC-QVCO. Figure 5.13a shows the circuit schematic used to analyze the quadrature-coupling transconductance, where DC bias circuitry and LC tanks are not shown as they do not affect the analysis.

The coupling path from qVCO to iVCO has been disconnected and the loading effect resulted from capacitor C_{cc} and C_{qc} is model by impedance Z_{b1} expressed as

$$Z_{b1} = \frac{1 + sR_b(C_{cc} + C_{qc})}{sC_{qc}(1 + sR_bC_{cc})} \tag{5.16}$$

In order to find the transconductance, the outputs of the second VCO are grounded and a differential voltage is applied to the iVCO as shown in Fig. 5.13a. The $G_c(s)$ can be found with the following expression:

$$G_c(s) = \frac{\Delta I_C}{\Delta V} = \frac{I_{C+} - I_{C-}}{\Delta V_+ - \Delta V_-} \tag{5.17}$$

The circuit in Fig. 5.13a is differential, and it can be simplified to half circuit as shown in Fig. 5.13b. The following two equations are defined to simplify the derivations:

$$Z_{b2} = \frac{1 + sR_b(C_{cc} + C_{qc})}{sC_{cc}(1 + sR_bC_{qc})} \tag{5.18}$$

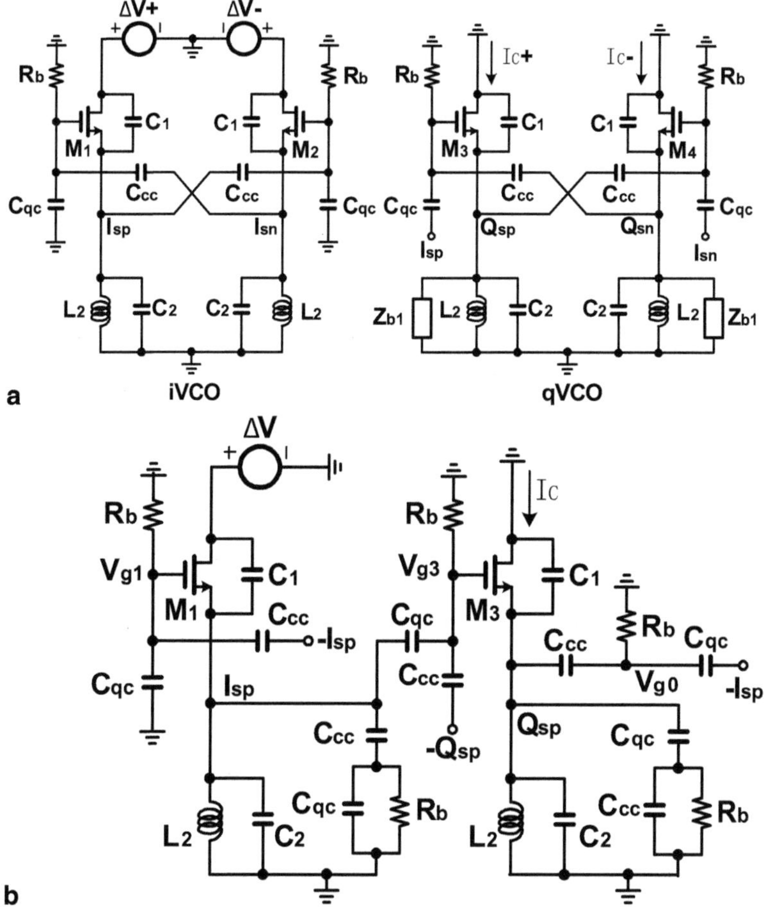

Fig. 5.13 a CC-QVCO circuit for the derivation of $G_c(s)$. **b** Simplified half-circuit model for the derivation of $G_c(s)$. The DC bias and varactors (included in C2) are not shown in the figure and the ground symbols represent AC ground

$$Z_2 = sL_2 \, \| \, \frac{1}{sC_2} \, \| \, R_{P2} \tag{5.19}$$

where R_{P2} represents the equivalent parallel resistance of the inductor L_2.

The voltage on the gate of M_1 is determined by the voltage at source terminal V_{isp} and it is given by

$$V_{g1} = -V_{isp} \frac{sR_b C_{cc}}{1 + sR_b (C_{cc} + C_{qc})} \tag{5.20}$$

Applying KCL at node I_{sp}, the currents flowing into the node is zero, namely,

$$g_{\mathrm{m}}(V_{\mathrm{g1}}-V_{\mathrm{isp}})+sC_1(\Delta V-V_{\mathrm{isp}})-\left(\frac{1}{Z_2}+\frac{1}{Z_{\mathrm{b2}}}\right)V_{\mathrm{isp}}-sC_{\mathrm{qc}}(V_{\mathrm{isp}}-V_{\mathrm{g2}})=0 \quad (5.21)$$

Similarly, the KCL equations for the node V_{g2}, Q_{sp}, V_{g0}, and output can be written as

$$sC_{\mathrm{qc}}(V_{\mathrm{isp}}-V_{\mathrm{g2}})-\frac{1}{R_{\mathrm{b}}}V_{\mathrm{g2}}-sC_{\mathrm{cc}}(V_{\mathrm{g2}}+V_{\mathrm{qsp}})=0 \quad (5.22)$$

$$I_C-\left(\frac{1}{Z_2}+\frac{1}{Z_{\mathrm{b1}}}\right)V_{\mathrm{qsp}}-sC_{\mathrm{cc}}(V_{\mathrm{qsp}}-V_{\mathrm{g0}})=0 \quad (5.23)$$

$$sC_{\mathrm{cc}}(V_{\mathrm{qsp}}-V_{\mathrm{g0}})-\frac{1}{R_{\mathrm{b}}}V_{\mathrm{g0}}-sC_{\mathrm{qc}}(V_{\mathrm{g0}}+V_{\mathrm{isp}})=0 \quad (5.24)$$

$$I_C-g_{\mathrm{m}}(V_{\mathrm{g2}}-V_{\mathrm{qsp}})+sC_1 V_{\mathrm{qsp}}=0 \quad (5.25)$$

Replacing the V_{g1} and V_{g2} with the value obtained from (20) and (22), respectively, Eq. (5.21) can be rewritten as:

$$b_0 V_{\mathrm{isp}}+b_1 V_{\mathrm{qsp}}=\Delta V \quad (5.26)$$

$$b_0=1+\frac{1}{sC_1 Z_2}+\frac{g_{\mathrm{m}}}{sC_1}+\frac{g_{\mathrm{m}}R_{\mathrm{b}}C_{\mathrm{cc}}+C_{\mathrm{cc}}+C_{\mathrm{qc}}+2sR_{\mathrm{b}}C_{\mathrm{cc}}C_{\mathrm{qc}}}{C_1\left[1+sR_{\mathrm{b}}\left(C_{\mathrm{cc}}+C_{\mathrm{qc}}\right)\right]} \quad (5.27)$$

$$b_1=\frac{sR_{\mathrm{b}}C_{\mathrm{cc}}C_{\mathrm{qc}}}{C_1\left[1+sR_{\mathrm{b}}\left(C_{\mathrm{cc}}+C_{\mathrm{qc}}\right)\right]} \quad (5.28)$$

From Eqs. (5.22) and (5.24), it can be seen that V_{g2} and V_{g0} are of opposite signs.

$$V_{\mathrm{g2}}=-V_{\mathrm{g0}}=\frac{sg_{\mathrm{m}}R_{\mathrm{b}}(C_{\mathrm{qc}}V_{\mathrm{isp}}-C_{\mathrm{cc}}V_{\mathrm{qsp}})}{1+sR_{\mathrm{b}}(C_{\mathrm{cc}}+C_{\mathrm{qc}})} \quad (5.29)$$

Combining Eqs. (5.23), (5.25), and (5.29), the solution for V_{isp} and V_{qsp} can be found as:

$$V_{\mathrm{isp}}=b_2 I_C=\left\{b_4+b_3\left[\frac{C_{\mathrm{cc}}}{C_{\mathrm{qc}}}+b_4\left(g_{\mathrm{m}}+sC_1\right)\right]\right\}I_C \quad (5.30)$$

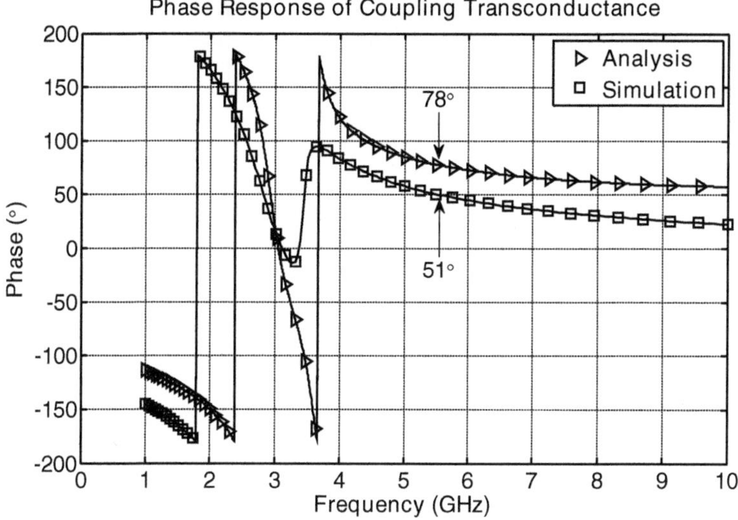

Fig. 5.14 Simulation results of phase shift versus analytical formula for quadrature-coupling transconductance

$$V_{qsp} = b_3 I_C = \frac{g_m - sC_{cc}}{g_m\left(\dfrac{1}{Z_2} + \dfrac{1}{Z_{b1}}\right) + sC_{cc}(2g_m + sC_1)} I_C \tag{5.31}$$

where b_2, b_3, and b_4 are used to simplify the expression and

$$b_4 = \frac{1 + sR_b(C_{cc} + C_{qc})}{sg_m R_b C_{cc}} \tag{5.32}$$

The relationship between ΔV and output current I_C can be solved by replacing the V_{isp} and V_{qsp} with Eqs. (5.30) and (5.31). Then, the final solution for the coupling transconductance of the proposed QVCO can be written as

$$G_C(s) = \frac{I_C}{\Delta V} = \frac{1}{b_0 b_2 + b_1 b_3} \tag{5.33}$$

In Fig. 5.14, we compare predictions from the above equations with simulations for the following component values: $C_1 = 0.8$ pF, $C_2 = 1.2$ pF, $L_2 = 1.25$ nH, $g_m = 10.3$ mS, $R_b = 200$ Ω, $C_{qc} = 95$ fF, $C_{cc} = 135$ fF, and $Q_{L2} = 15$. As the analytical equation does not include parasitic capacitances and resistances such as C_{GS}, C_{GD}, and r_{DS}, the simulated phase shift of 51° is lower than the one obtained from the calculation of 78° at 5.6-GHz frequency. According to the simulation done in MMSIM 10.2, the parasitic capacitance C_{GS} is found to be 50 fF which is comparable to the cross-

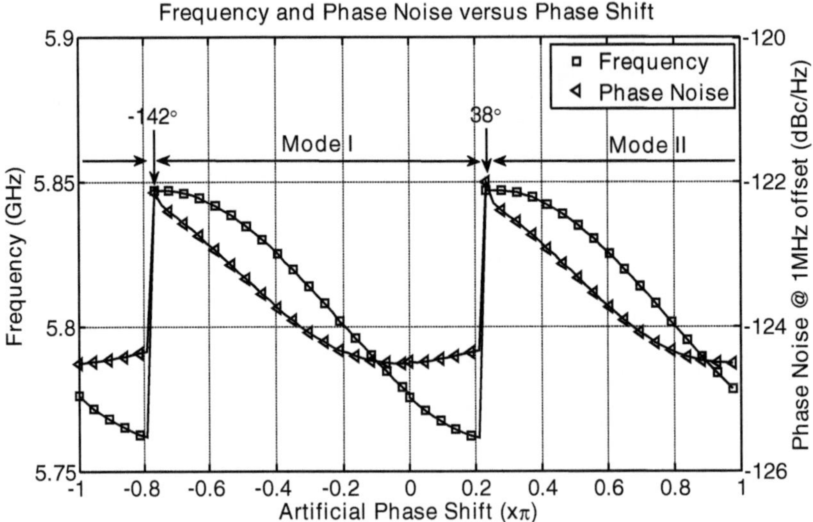

Fig. 5.15 Simulation results of oscillation frequency and phase noise with artificial phase shift introduced in the coupling path

coupled capacitor C_{cc} and quadrature-coupling capacitor C_{qc} and it explains the difference. Nevertheless, the analytical derivation can help predicting the phase shift available to avoid ambiguous oscillation.

In order to illustrate that the proposed CC-QVCO can operate at only one of the two stable modes, simulations are done with artificial phase shifts introduced into the quadrature-coupling path. The simulation results are obtained from the PSS plus P-noise feature of MMSIM 10.2. Figure 5.15 shows the phase noise and the frequency variation with artificial phase shift, where the separation region of mode I and mode II is at 38° and − 142°. When the artificial phase shift is between − 142° and 38°, the output phase of i-VCO lags that of the qVCO by 90°, which is defined as mode I. For mode II, the output phase of iVCO leads qVCO by 90°. Without the 38°-phase margin, the region separating mode I and mode II would be around 0°. In other words, the intrinsic phase shift of the CC-QVCO has already shifted the separating region from 0° to 38°. Since the result of 38° is obtained from large signal simulation, the practical safety margin for phase shift has been dropped down compared with the ac simulation result of 51°. Moreover, the phase noise is optimum around 0°-artificial phase shift as shown in Fig. 5.15. Therefore, the proposed CC-QVCO achieves good phase noise performance with the intrinsic phase shift of 38° introduced by the coupling path. On the other hand, the variations of the simulated phase noise is less than 2.5 dB, that is to say, the phase noise performance is not so sensitive to the variation of phase shift and the phase shift can be increased to leave more margin for stable oscillation.

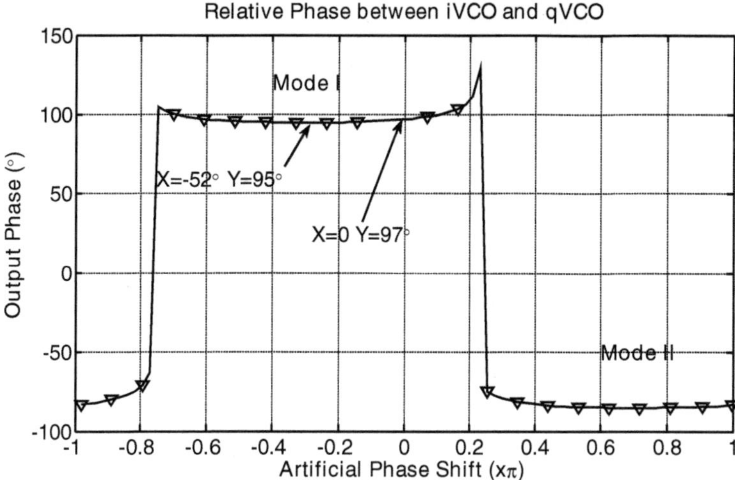

Fig. 5.16 Simulation result of output phases with artificial phase shift introduced in the coupling path (with $C_{tankq} = 1.01 C_{tanki}$)

5.2.6 Quadrature Inaccuracy

Mismatches between the two VCO cores for quadrature generation cause the outputs to deviate from $\pm 90°$ condition and the amplitudes to be unequal. Compared with unequal amplitude, phase accuracy between the quadrature outputs is the primary concern for QVCO circuits since the amplitude mismatch has less impact on receiving or transmitting mixers if limiting buffers are used. Conventional QVCO with parallel coupling is less sensitive to component mismatch when the introduced phase shift in the coupling path is around 90° [10–12]. According to this theory, the proposed CC-QVCO would show minimum sensitivity around $-52°$ artificial phase shift since the intrinsic phase shift in the coupling path is 38°. Figure 5.16 shows the simulated quadrature inaccuracy due to 1 % mismatch between the resonant LC tanks. A phase shift of 90° in the coupling path is also the optimum point for the CC-QVCO to reduce the phase inaccuracy arising from component mismatches. The simulation results further validate the theory proposed by [10]. Although not optimum, the intrinsic phase shift of 38° in the coupling path is good enough to avoid ambiguous oscillation.

5.3 Implementation and Measured Results

The CC-QVCO was implemented in a 0.13 μm CMOS technology and the die photo of the chip is shown in Fig. 5.17. The QVCO core including the pads and testing output buffers occupies an area of $1.2 \times 1.2 mm^2$, while the core of the QVCO

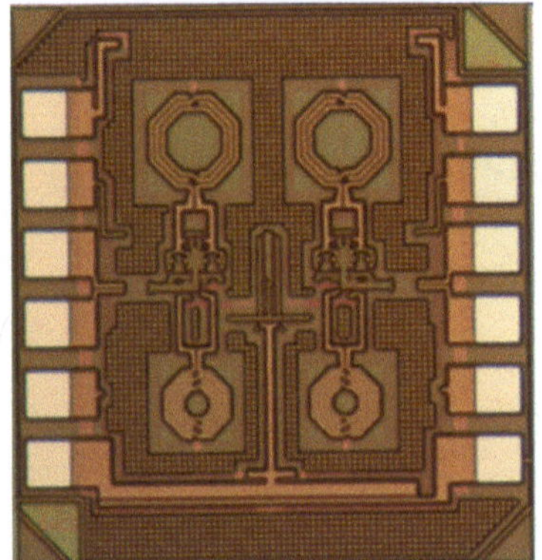

Fig. 5.17 Die photo of the implemented QVCO RFIC (1.2×1.2 mm^2 including pads)

takes only 0.6×0.8mm^2. After careful tradeoff between phase noise improvement and phase accuracy, a quadrature-coupling strength factor of $m = 0.4$ is chosen with $C_{cc} = 95$ fF and $C_{qc} = 135$ fF. Both the load tank L_1 and L_2 are of symmetric structure and their values are: $L_{1,\text{diff}} = 2.05$ nH and $L_{2,\text{diff}} = 2.5$ nH. The extracted parasitic capacitance at each source terminal is around 300 fF mainly caused by the cross-coupled wiring and diode wiring. Therefore, the output frequency of the CC-QVCO is more stable with the cross-coupled connections and the varactors placing at the source terminal than at the drain terminal.

The phase noise is measured using an Agilent E4446A spectrum analyzer with phase noise option. The design provides the reconfigurability to form either a CC-QVCO or a SVCO for comparison. Figure 5.18 shows the phase noise for both the CC-QVCO and SVCO measured under the same bias conditions. The CC-QVCO

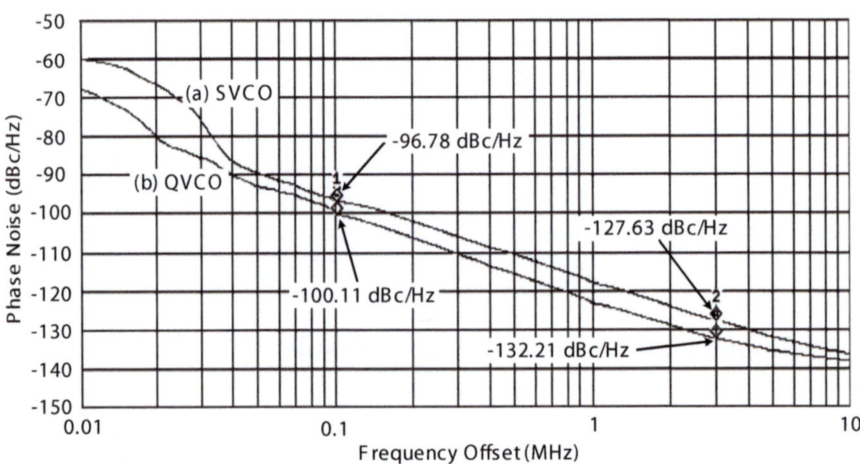

Fig. 5.18 Measured phase noise of **a** SVCO, and **b** QVCO

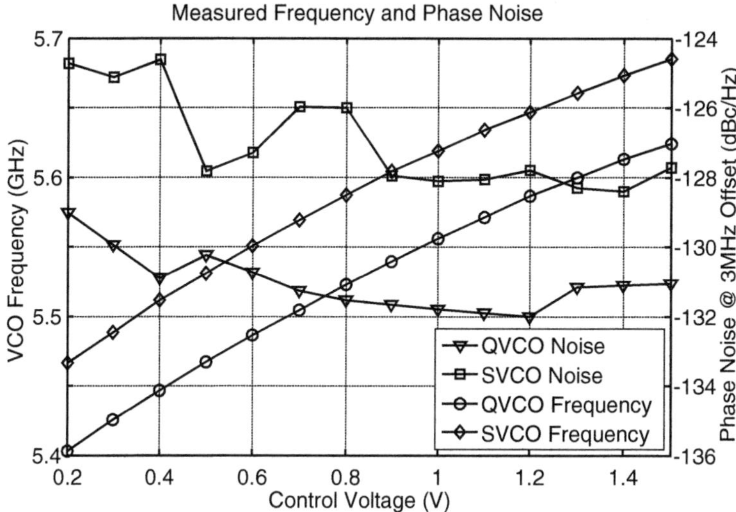

Fig. 5.19 Measured frequency tuning range and phase noise of QVCO and SVCO

and SVCO achieved measured phase noise of -132.2 dBc/Hz and -127.6 dBc/Hz @ 3-MHz offset while consuming 4.2 mW and 2.1 mW, respectively. The phase noise improvement of the CC-QVCO over SVCO is about 3.3 to 4.6 dB from 100-kHz to 10-MHz offset frequency range. The measurement result demonstrates the effectiveness of the proposed capacitive coupling in improving the phase noise performance. The measured FoMs at 3-MHz offset are 190 and 191.4 dB for SVCO and QVCO, respectively.

The frequency tuning ranges from 5.4 GHz to 5.62 GHz for CC-QVCO and from 5.46 GHz to 5.68 GHz for SVCO, respectively, as shown in Fig. 5.19. The phase noise of QVCO varies from -129.5 dBc/Hz to -132.2 dBc/Hz @ 3-MHz offset in the entire tuning frequency range. The measured phase noises for both the QVCO and SVCO are about $2.7 \sim 4$ dB higher than the simulation results across the tuning range; but the measured noise improvement of QVCO over its SVCO counterpart agrees well with the simulation results. A larger tuning range can be achieved by including additional digital controlled capacitor array in parallel with the load tank or the bottom tank. The phase noise for both the QVCO and SVCO with larger tuning range will increase; and the noise degradation depends on the quality factor of the additional capacitor array or varactor. However, the phase noise improvement for the QVCO compared with its SVCO core is still valid.

Since the CC-QVCO includes a second inductor L_2 at the bottom to enhance the signal swing, this second tank might start oscillation and generate unwanted second oscillation frequency. The resonant frequency of the bottom tank has been kept about 2.5 GHz below that of the load tank. Figure 5.20 shows the measured output spectrum at 5.6 GHz frequency. The output spectrum is clean without any unwanted resonant frequency around 3 GHz that might be generated by the bottom tank.

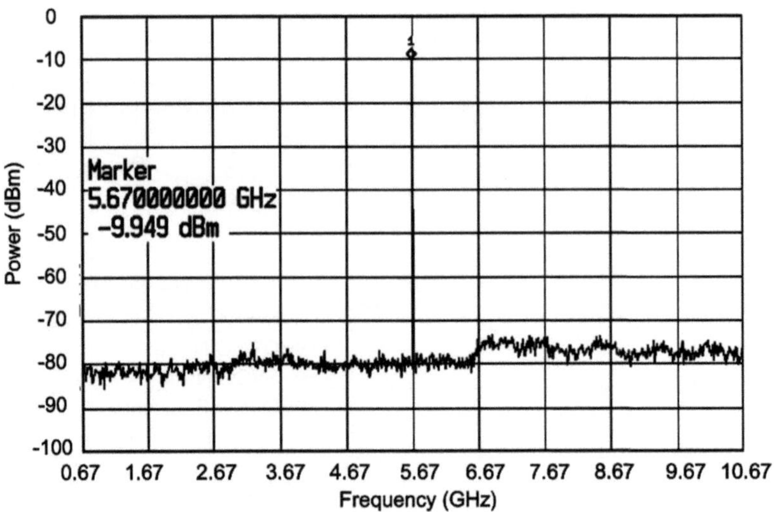

Fig. 5.20 Measured output spectrum for the CC-QVCO

The output voltage waveforms at 5.6 GHz frequency are shown in Fig. 5.21. As being described in Sect. II.-E, the intrinsic phase shift of 38° in the coupling path helps the proposed CC-QVCO to generate quadrature outputs of +90° between iVCO and qVCO. Three prototype boards have been measured to observe the output voltage and the quadrature accuracy. The quadrature phase error ranges from 0.7~3° for the three prototypes. Assuming all the phase error is caused by the tank mismatch, this measured phase error corresponds to a tank mismatch less than 1 %. Assuming this phase error is the only error existing in an image rejection receiver,

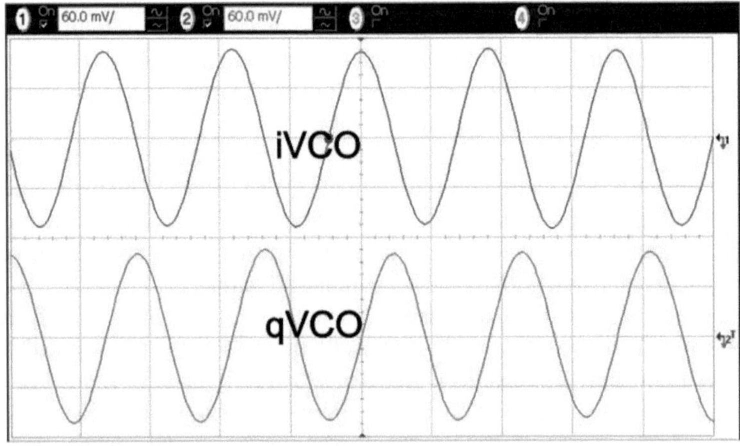

Fig. 5.21 Measured output voltage waveforms for the CC-QVCO

this phase error would cause to an image rejection ratio less than 31.6 dB. It should
be pointed out that the measured phase error is also contributed by the mismatches
between the quadrature signal paths including the buffers, package pins/pads, ca-
bles, and connectors. The actual phase error caused by the QVCO should be smaller
than the error observed. Moreover, the quadrature relationships between iVCO and
qVCO are deterministic for the three measured boards. Therefore, the phenomenon
of bimodal oscillation has been avoided for the three prototypes.

Table 5.1 summarizes the performance of the proposed CC-QVCO and comparison with previously published QVCO work. When compared with the prior art, the
proposed Colpitts QVCO achieves a FoM of 191.4 dB, where the FoM and FOM_T
are defined as [24]:

$$\text{FoM} = 10\log\left[\left(\frac{f_0}{\Delta f}\right)^2 \frac{1\text{mW}}{P}\right] - L\left(\Delta f\right) \tag{5.34}$$

$$\text{FoM}_\text{T} = \text{FoM} + 10\log\left[\frac{\text{TR}\left(\%\right)}{\text{Vtune}}\right] \tag{5.35}$$

In the above equations, $L(\Delta f)$ is the phase noise at the Δf offset from the oscillator
frequency f_0, P is the QVCO's core power consumption in mW, TR is the relative
tuning range, and Vtune is the corresponding range of tuning voltage.

5.4   Conclusions

A CMOS enhance-swing Colpitts QVCO with capacitive coupling (CC-QVCO)
for noise reduction is proposed and analyzed in this chapter. The prototype CMOS
CC-QVCO was fabricated in 0.13 µm CMOS technology with measured frequency tuning range about 4 %. The CC-QVCO achieves an FoM of 191.4 dB while
the FoM of a SVCO of the same type is 190 dB. The phase noise improvement
over SVCO is 3.3 and 4.6 dB at 100-kHz and 3-MHz offset, respectively. Moreover, the intrinsic phase shift in the quadrature-coupling path has been analyzed,
showing advantages of avoiding bimodal oscillations for QVCO operations. The
measurement results demonstrate not only the effectiveness of the noise improvement using the proposed optimized capacitive-coupling technique, but also the
intrinsic phase shift that improves the stability of the CC-QVCO. The CC-QVCO
consumes only 4.2-mW power with a 0.6-V supply and occupies a core area of
0.48 mm².

Table 5.1 Performance summary and comparison of QVCOs with different coupling techniques

	[1]	[2]	[4]	[8]	[9]	[10]	[12]	This work
Coupling mechanism	Super harmonic coupling	Series coupling at source	Series coupling at drain	Energy circulating	Transformer coupling	Capacitive source coupling	Capacitor coupling	CC-QVCO
Frequency (GHz)	4.88	2.2	2	5.3	17	2.07	10.4	5.6
Power (mW)	22	8.6	20.8	20.7	5	4.5	3.6	4.2
Power supply (V)	2.5	2.0	1.3	1.8	1.0	1.5	1.5	0.6
Phase noise@1 MHz (dBc/Hz)	-125	-127	$-140@3M$	-134.4	-110	-124.4	-115.7	-122 -132.2 @3M
Phase accuracy (°)	2.6	–	0.6	–	1.4	–	1.5	3
Technology	0.25 µm CMOS	0.18 µm CMOS	0.35 µm CMOS	0.18 µm CMOS	0.18 µm BiCMOS	0.25 µm BiCMOS	0.18 µm CMOS	0.13 µm CMOS
Area (mm^2)	–	1.36	1.26	3.04	0.126 (core)	0.625	0.77	0.48 (core)
Tuning range	12%	20%	17%	1%	16.5%	18%	10%	4%
FOM (dB)	185	184.5	183.3	196	187.6	186	190.5	191.4
FOM$_T$ (dB)	191.8	194.5	194.5	193.4	199.8	196.8	198.7	196

References

1. S. L. J. Gierkink, S. Levantino, R. C. Frye, C. Samori, and V. Boccuzzi, "A low phase-noise 5-GHz CMOS quadrature VCO using superharmonic coupling," IEEE J. Solid-State Circuits, vol. 38, no. 7, pp. 1148–1154, Jul. 2003.
2. C. Yao and A. N. Willson, "A phase-noise reduction technique for quadrature LC-VCO with phase-to-amplitude noise conversion," in IEEE Int. Solid-State Circuits Conf. Dig. Tech. Papers, Feb. 2006, pp. 701–710.
3. A. W. L. Ng and H. C. Luong, "A 1-V 17-GHz 5-mW CMOS quadrature VCO based on transformer coupling," IEEE J. Solid-State Circuits, vol. 42, no. 9, pp. 1933–1941, Sep. 2007.
4. B. Soltanian and P. Kinget, "A low phase noise quadrature LC VCO using capacitive common-source coupling," in Proc. European Solid-State Circuits Conf., Sep. 2006, pp. 436–439.
5. C. T. Fu and H. C. Luong, "A 0.8-V CMOS quadrature LC VCO using capacitive coupling," in IEEE Asian Solid-State Circuits Conf., Nov. 2007, pp. 436–439.
6. I. Shen and C. F. Jou, "A X-band capacitor-coupled QVCO using sinusoidal current bias technique," IEEE Trans. Microw. Thoery Tech., vol. 60, no. 2, pp. 318–328, Feb. 2012.
7. B. Razavi, RF Microelectronics Second Edition, New Jersey, USA, Prentice Hall, 2011
8. S. Li, I. Kipnis, and M. Ismail, " A 10-GHz CMOS quadrature LC-VCO for multirate optical applications," IEEE J. Solid-State Circuits, vol. 38, no. 10, pp. 1626–1634, Oct. 2003.
9. H. Tong, S. Cheng, Y. Lo, A. I. Karsilayan, and J. Silva-Martinez, "An LC quadrature VCO using capacitive source degeneration coupling to eliminate bi-modal oscillation," IEEE Trans. Circuits Syst. I, Reg. Papers, vol. 59, no. 10, pp. 1–9, Oct. 2012.
10. A. Mirzaei, M. E. Heidari, R. Bagheri, S. Chehrazi, and A. A. Abidi, "The quadrature LC oscillator: a complete portrait based on injection locking," IEEE J. Solid-State Circuits, vol. 42, no. 9, pp. 1916–1932, Sep. 2007.
11. J. van der Tang, P. van de Ven, D. Kasperkoviz, and A. van Roermund, "Analysis and design of an optimally coupled 5-GHz quadrature LC oscillator," IEEE J. Solid-State Circuits, vol. 37, no. 5, pp. 657–661, May 2002.
12. D. Huang, W. Li, J. Zhou, N. Li, and J. Chen, "A frequency synthesizer with optimally coupled QVCO and harmonic-rejection SSB mixer for multi-standard wireless receiver," IEEE J. Solid-State Circuits, vol. 46, no. 6, pp. 1307–1320, June 2011.
13. A. Hajimiri and T. H. Lee, "A general theory of phase noise in electrical oscillators," IEEE J. Solid-State Circuits, vol. 33, no. 2, pp. 179–194, Feb. 1992.
14. F. Zhao and F. F. Dai, "A 0.6V quadrature VCO with optimized capacitive coupling for phase noise reduction," in Proc. IEEE Custom Integrated Circuits Conf., Oct. 2011, pp. 1–4.
15. T. W. Brown, F. Farhabakhshian, A. G. Roy, T. S. Fiez, and K. Mayaram, "A 475 mV, 4.9 GHz enhanced swing differential Colpitts VCO with phase noise of -136 dBc/Hz at a 3 MHz offset frequency," IEEE J. Solid-State Circuits, vol. 46, no. 8, pp. 1782–1795, Dec. 2011.
16. J. W. M. Rogers, J. A. Macedo, and C. Plett, "The effect of varactor nonlinearity on the phase noise of complementely integrated VCOs," IEEE J. Solid-State Circuits, vol. 35, no. 9, pp. 1360–1367, Sep. 2000.
17. J. W. M. Rogers, F. F. Dai, M. S. Cavin, and D. G. Rahn, "A multiband $\Delta\Sigma$ fractional-N frequency synthesizer for a MIMO WLAN transceiver RFIC," IEEE J. Solid-State Circuits, vol. 40, no. 3, pp. 678–689, Mar. 2005.
18. X. Li, S. Shekhar, and D. J. Allstot, "Gm-boosted common-gate LNA and differential Colpitts VCO/QVCO in 0.18-μm CMOS," IEEE J. Solid-State Circuits, vol. 40, no. 12, pp. 2609–2619, Dec. 2005.
19. P. Andreani and X. Wang, "On the phase-noise and phase-error performances of multiphase LC CMOS VCOs," IEEE J. Solid-State Circuits, vol. 39, no. 11, pp. 1883–1893, Nov. 2004.
20. P. Andreani, "A 2 GHz, 17 % tuning range quadrature CMOS VCO with high figure-of-merit and 0.6° phase error," in Proc. European Solid-State Circuits Conf., Aug. 2002, pp. 815–818.

21. A. Rofougaran, J. Rael, M. Rofougaran, and A. Abidi, "A 900 MHz CMOS LC-oscillator with quadrature outputs," in IEEE Int. Solid-State Circuits Conf. Dig. Tech. Papers, Feb. 1996, pp. 392–393.
22. T. Liu, "A 6.5-GHz monolithic CMOS voltage-controlled oscillator," in IEEE Int. Solid-State Circuits Conf. Dig. Tech. Papers, Feb. 1999, pp. 404–405.
23. P. Andreani, A. Bonfanti, L. Romano, and C. Samori, "Analysis and design of a 1.8-GHz CMOS LC quadrature VCO," IEEE J. Solid-State Circuits, vol. 37, no. 12, pp. 1737–1747, Dec. 2002.
24. B. D. Muer, N. Itoh, M. Borremans, and M. Steyaert, "A 1.8 GHz highly-tunable low-phase noise CMOS VCO," in Proc. IEEE Custom Integr. Circuits Conf., Oct. 2000, pp. 585–588.

Chapter 6
Conclusions

This book presents detailed noise analysis and quantization noise suppressing techniques for fractional-phase-locked loop (PLL) clock generation systems. Also analyzed is the nonlinearity impact on the sigma-delta modulator-based PLL frequency synthesizer. Several techniques for phase noise reduction have been discussed and a noise reduction technique for higher order SDM is proposed. Moreover, a noise behavioral model including the charge pump nonlinearity is developed for fractional-N PLL.

The next topic of this book is a wide-band integer-N PLL for wireless transceiver design. Detailed theoretical analysis of a Bi-CMOS bandgap is given to show the improved PSRR over its conventional counterpart who uses bipolar transistor only. Simple design methodology for power optimization of dividers has been proposed and proved with simulation results.

Then capacitive-coupling technique is introduced for multiphase clock generation and QVCOs to achieve the goal of improved phase noise performance and elimination of bi-modal oscillation. A key contribution of this work was the capacitive-coupling technique for quadrature generation which is free from the problem of the noise degradation in the quadrature-coupling path and phase ambiguity. Meantime, it shows noise improvement over its single-phase counterpart. Theoretical analysis, combined with simulation and experimental results, is given to show the efficiency of the proposed quadrature-coupling technique.

A 0.6 V QVCO prototype is implemented in 0.13 μm CMOS technology. Differential Colpitts QVCO with capacitive-coupling technique and enhanced swing has been optimized to achieve minimum phase noise performance. The capacitive-coupling ratio can be chosen for either better phase noise or phase error performance to meet specs for different applications. Different from classic QVCOs using active parallel coupling technique, the capacitive coupling provides better phase noise performance due to its improved ISF. A secondary inductor is inserted at the bottom to enhance the output signal swing, allowing improved phase noise performance at very low supply voltage.

For the most popular commercial applications, QVCO with well-controlled current bias is preferred due to its advantage of simple structure and high reliability.

© Springer International Publishing Switzerland 2015

F. Zhao, F. F. Dai, *Low-Noise Low-Power Design for Phase-Locked Loops,*

DOI 10.1007/978-3-319-12200-7_6

Another QVCO structure with capacitive-coupling is also introduced to provide robust solution for such applications. This CC-QVCO shows excellent phase noise tolerance and phase error over wide frequency tuning range. Silicon implementation and measurement are given to verify the proposed quadrature-coupling technique.

The proposed capacitive coupling technique is very suitable to implement a quadrature oscillator with improved phase noise performance. The phase delay introduced to avoid phase ambiguity is around 40° for the 0.6-V Colpitts QVCO and hard to control, especially at frequencies above 10 GHz because of the short timeperiod. One future direction might be to develop phase delaying generation technique that is suitable for frequency above 10 GHz applications. Moreover, the capacitive coupling technique is also suitable for multiphase oscillator, such as three or four phase. Another promising direction is to develop multiphase VCO with the proposed capacitive-coupling technique for phase array transceivers due to its promising feature of noise reduction.

The proposed quantization noise reduction technique is useful and efficient. But further system level simulation and circuit design is not included. Therefore, one possible future work is to design a fractional-N PLL with artificially introduced nonlinearity and then verify the model through silicon verification.